Pitman Research Notes in Mathematics Series

Main Editors
H. Brezis, Université de Paris
R. G. Douglas, State University of New York at Stony Brook
A. Jeffrey, University of Newcastle-upon-Tyne *(Founding Editor)*

Editorial Board
R. Aris, University of Minnesota
A. Bensoussan, INRIA, France
S. Bloch, University of Chicago
B. Bollobás, University of Cambridge
W. Bürger, Universität Karlsruhe
S. Donaldson, University of Oxford
J. Douglas Jr, University of Chicago
R. J. Elliott, University of Alberta
G. Fichera, Università di Roma
R. P. Gilbert, University of Delaware
R. Glowinski, Université de Paris
K. P. Hadeler, Universität Tübingen
K. Kirchgässner, Universität Stuttgart
B. Lawson, State University of New York at Stony Brook
W. F. Lucas, Claremont Graduate School
R. E. Meyer, University of Wisconsin-Madison
S. Mori, Nagoya University
L. E. Payne, Cornell University
G. F. Roach, University of Strathclyde
J. H. Seinfeld, California Institute of Technology
B. Simon, California Institute of Technology
I. N. Stewart, University of Warwick
S. J. Taylor, University of Virginia

Submission of proposals for consideration

Suggestions for publication, in the form of outlines and representative samples, are invited by the Editorial Board for assessment. Intending authors should approach one of the main editors or another member of the Editorial Board, citing the relevant AMS subject classifications. Alternatively, outlines may be sent directly to the publisher's offices. Refereeing is by members of the board and other mathematical authorities in the topic concerned, throughout the world.

Preparation of accepted manuscripts

On acceptance of a proposal, the publisher will supply full instructions for the preparation of manuscripts in a form suitable for direct photo-lithographic reproduction. Specially printed grid sheets are provided and a contribution is offered by the publisher towards the cost of typing. Word processor output, subject to the publisher's approval, is also acceptable.

Illustrations should be prepared by the authors, ready for direct reproduction without further improvement. The use of hand-drawn symbols should be avoided wherever possible, in order to maintain maximum clarity of the text.

The publisher will be pleased to give any guidance necessary during the preparation of a typescript, and will be happy to answer any queries.

Important note

In order to avoid later retyping, intending authors are strongly urged not to begin final preparation of a typescript before receiving the publisher's guidelines and special paper. In this way it is hoped to preserve the uniform appearance of the series.

Longman Scientific & Technical
Longman House
Burnt Mill
Harlow, Essex, UK
(tel (0279) 426721)

Invariant Manifold Theory For Hydrodynamic Transition

S S Sritharan

University of Southern California

Invariant Manifold Theory for Hydrodynamic Transition

Longman
Scientific &
Technical

Copublished in the Sonited States wit
John Wiley & Sons, Inc., New York

Longman Scientific & Technical,
Longman Group UK Limited,
Longman House, Burnt Mill, Harlow
Essex CM20 2JE, England
and Associated Companies throughout the world.

Copublished in the United States with
John Wiley & Sons, Inc., 605 Third Avenue, New York, NY 10158

First published 1990

AMS Subject Classification: 76D05, 76F99, 76E30, 35B32, 35B37, 35Q10, 58F15, 58F14, 58F10

ISSN 0269-3674

British Library Cataloguing in Publication Data
Sritharan, S. S. (Sivaguru Sornalingam) 1954–
    Invariant manifold theory for hydrodydnamic transition.
        1. Topological spaces. Dynamical systems
    I. Title
    514.3

ISBN 0-582-06781-2

Library of Congress Cataloging-in-Publication Data
Sritharan, S. S. (Sivaguru Sornalingam), 1954–
    Invariant manifold theory for hydrodynamic transition / S. S. Sritharan.
        p. cm. — (Pitman research notes in mathematics series, ISSN 0269-3674; 241)
        Includes index.
        1. Navier-Stokes equations. 2. Turbulence. 3. Chaotic behavior in systems. 4. Bifurcation
    theory. 5. Manifolds. 6. Invariants.
    I. Title. II. Series.
    QA929.S75   1990
    537'.516'0151607—dc20   90-13359
                                                                          CIP

Printed and bound in Great Britain
by Biddles Ltd, Guildford and King's Lynn

Contents

Preface

Author wishes to thank professor Klaus Kirchgassner for kindly reviewing the manuscript. Initial stages of this research were supported by the Faculty research innovation fund of the University of Southern California. Main funding was provided by the Office of Naval Research under the URI-Contract No. N00014-86-K-0679 supervised by Drs. Mike Reischman and Spiro Lekoudis. Author would like to thank for their understanding and patience.

S. S. Sritharan,

Department of Aerospace Engineering,

University of Southern California,

Los Angeles, California 90089-1191,

U. S. A

March 1990.

Chapter 1

Introduction

An invariant manifold theory for the Navier-Stokes equations would lay a bridge between the theory of finite dimensional dynamical systems and the onset dynamics of turbulence. In this work we will develop such a theory for hydrodynamic motions in bounded containers. Although the main focus of this work is on bounded domains we will discuss extensions to unbounded domains in each section and point out open problems. *In simple words, invariant manifold theory identifies the active and slave modes of a particular solution orbit with respect to a basic solution orbit.*

It has been proven by Ladyzhenkaya[47] that the long time behavior of viscous flows in two dimensional bounded domains can be characterized by a compact attractor. This attractor contains (atleast) all the stationary, time periodic and quasiperiodic solutions. Moreover, the Navier-Stokes equations define a dynamical system in this attractor. This means that the solution (in this attractor) is smooth and defined for positive as well as negative times. Numerical approximations and mathematical regularizations of the Navier -Stokes equations produce upper semicontinuous global attractors [41, 66] which converge to the global attractor of the conventional system as the approximation (or the regularization) parameter approaches zero. These nice properties of the global attractors make them the central object of research for turbulence theory. It is also of interest to study the orbits nearby such attractors.

For the case of three dimensional flows, existence of a compact global attractor has not been proven yet. As one might expect, this problem is connected with the global unique solvability theorem which is not available for the three dimensional case. Hence, the fundamental step in the dynamical systems theory for the Navier-Stokes equations is the global unique solvability theorem which is, in some sense, same as the continuous dependence theorem. The Hadamard's notion of wellposedness requires the existence, uniqueness and continuous dependence of the solution with respect to data. Unfortunately a complete answer to this basic question is available only for two dimensional (time dependent) viscous flow in bounded domains and for certain unbounded flow problems. For these problems it is possible to prove that the solution exists, unique, holomorphic in time in the neighborhood of the positive real axis and is Frechet analytic in initial and boundary data. The solvability problem and hence the dynamical systems theoretical description is not complete for the other cases. *The global unique solvability problem for viscous flow in three dimensional bounded (or unbounded) domains is now regarded as one of the profound open problems in mathematical science.* In the simplest setting this problem can be described as follows. Consider the viscous flow in a three dimensional bounded container with homogeneous (nonslip) boundary conditions. Motion is initiated by prescribing an initial velocity distribution with finite but arbitrary amount of energy. In other words the initial velocity has finite $L^2(\Omega)$ norm. For this case there is a famous theorem due to Hopf[38] which provides the existence of a weak solution with the following properties. The energy $\mathcal{E}(t)$ or the $L^2(\Omega)$ norm of the velocity field is bounded for all subsequent times $(\mathcal{E}(t) \leq \mathcal{E}(0))$ and the enstrophy $\mathcal{E}_N(t)$ which is the square integral of the vorticity (also the Dirichlet integral of the velocity) is square integrable in time. Unfortunately this information is not sufficient to prove the uniqueness of this class of weak solutions. In two dimensions, however, this is in fact sufficient as demonstrated by Lions and Prodi[55]. For the two dimensional case of this problem, we can show that the solution is holomorphic in the neighborhood of the positive time axis and Frechet analytic in the initial data [89]. For the three dimensional case it is possible to establish the uniqueness of the Hopf class solutions if it is known that the fourth power of the enstrophy $\mathcal{E}_N(t)$ is integrable. In other words the $L^4(0,T)$ norm of $\mathcal{E}_N(t)$ is bounded. Is this possible ?. Since this additional requirement is expressed in terms of the vorticity in the flow, it is certainly reasonable to compare the nature of vorticity dynamics in two and three dimensions. In three dimensional flows there are additional features such as vortex stretching and knottedness of vortex filaments (which is associated with nontrivial helicity[62]). Hence, the resolution of the global unique solvability theorem may require a detailed understanding of the solutions of the three dimensional Euler equations.

Let us now formulate a slightly different problem. We consider again the three dimensional viscous flow problem with homogeneous boundary data and an initial velocity with finite energy and enstrophy. This time we will introduce a forcing (perhaps a distributed forcing in the sense of control theory) at the right hand side of the momentum equations. Here again, if the force is square integrable in space and time then we have a Hopf class weak solution with the same properties as above. In this case however Fursikov[24] has shown that there is a set of such forces (included in the class forces that are divergence free and square integrable in space and time) for which the additional requirement $\mathcal{E}_N(t) \in L^4(0,T)$ is satisfied. For this case the problem is uniquely solvable. *It is however not known whether this class of forces includes zero.* The Fursikov theorem however, is a generic solvability theorem. Thus in order to resolve the three dimensional homogeneous Navier Stokes problem we only need to introduce a force which is arbitrarily close to zero in a suitable topology.

In the light of Fursikov's result we may say that a possible method of resolving the global unique solvability problem is to show that his class of forces includes zero as an element. Mathematical regularizations [64, 65, 66] seem to provide another promising method. Here we add (to the momentum equations) higher order terms such as Laplacian square with artificial viscosity coefficients. For such system, the nonlinearity (the inertia term) exhibits much better behavior and allows us to establish global unique solvability up to several dimensions. The task then would be to show that, as the regularization parameter approaches zero, the solution of the regularized system converges in some topology to the solution of the conventional system. This has been accomplished for the case of low Reynold's numbers for three dimensional problems and for arbitrary Reynold's numbers for two dimensional problems.

In this monograph however, we will develop a dynamical systems theory about a smooth basic solution and hence local solvability theorems are sufficient for most purposes. We will consider stationary and time periodic smooth solutions and study the nearby solutions. The Reynold's number is held fixed and hence does not play an explicit role.

In the second chapter we will present the functional framework used in this monograph. Analysis of the Stokes problem with various boundary conditions is given. We define an accretive selfadjoint operator called the Stokes operator and characterize its fractional powers. The central result used here is the Cattabriga regularity theorem. Functional framework for unbounded domains are discussed along with a general theory of the Stokes operator.

In the third chapter we study the linearizations (of the stationary Navier-Stokes equations) about a stationary basic solution. We will also provide the existence and regularity results for stationary basic solutions. We then consider the linearized operator about a smooth basic field and establish its spectral properties. Although the spectral theorem 3.6 is stated for the case of stationary basic flow in fact it applies (for each time t) if the basic flow is time dependent. The completeness theorem 3.6 for the eigenfunctions is needed in any constructive procedure that uses the invariant manifold method to compute the bifurcating solutions. Mathematical theory of Exterior hydrodynamics including spectral theory is also discussed.

In chapter four we study the linear evolution problem obtained by linearizing the Navier-Stokes equations about a smooth time dependent basic flow. The main results in this section are the regularity and spectral properties of the monodromy operator. We also provide existence theorems for time periodic solutions.

The nonlinear semigroup associated with the Navier-Stokes equations is considered in chapter five. We establish the analyticity of the solution map with respect to the initial data. This property of the solution map enables us to establish the analyticity of the invariant manifolds in chapter seven. For three dimensional flows, because of the lack of global unique solvability theorem, the semigroup can be defined only up to a finite maximal time determined by the size of the initial data. However, it is possible to take a small enough initial data (in certain norm) so that the maximal time is infinity. For two dimensional flows of course the maximal time is infinity for arbitrary initial data.

In chapter six we establish the principle of linearized stability for stationary and time periodic basic solutions. Our result is independent of a specific topology. The Lyapunov instability is characterized using the *invariant cone* concept introduced by Kirchgassner and Scheurle [43] .

In chapter seven we refine this result and establish the existence and uniqueness of invariant manifolds. The central result of this chapter is the analyticity of these manifolds. We prove this result by working with classes of analytic maps throughout this section. The analyticity theorem provides us with an analytic coordinate chart for the unstable manifold and a convergent expansion method to compute bifurcating solutions.

In Appendix A, we present Ladyzhenskaya's theory of global attractors for two dimensional viscous flows in bounded domains.

In Appendix B, we will derive certain implications of group action on the invariant manifold theory.

Chapter 2

The Governing Equations And The Functional Framework

We will first formulate the mathematical problem of time dependent viscous flow in bounded containers with prescribed boundary velocity distributions. Formulations for other problems would require additional conditions such as farfield and flux conditions and they will be discussed later in this chapter. Let $\Omega \subset \boldsymbol{R}^n, n = 2$ or 3 be an open bounded set of class $C^r, r \geq 2$. The smoothness requirement here is not crucial to the developments of this monograph. We will consider the motion of a viscous incompressible fluid in Ω with the prescribed initial velocity field $\boldsymbol{u}_0$ and a time dependent boundary distribution $\boldsymbol{u}_b$. Let $\boldsymbol{u}, p$ and ν, respectively denote the velocity field, pressure field and kinematic viscosity. The problem is to find $(\boldsymbol{u}, p) : \Omega \times (0, \infty) \to R^n \times R$ such that,

$$\boldsymbol{u}_t + (\boldsymbol{u} \cdot \nabla)\boldsymbol{u} = -\nabla p + \nu \Delta \boldsymbol{u} \text{ in } \Omega \times (0, \infty)$$

$$\nabla \cdot \boldsymbol{u} = 0 \text{ in } \Omega \times (0, \infty) \tag{2.1}$$

$$\boldsymbol{u}(x, t) = \boldsymbol{u}_b(x, t) \text{ for } (x, t) \in \partial\Omega \times [0, \infty) \text{ with}$$

$$\int_{\partial\Omega} \boldsymbol{u}_b \cdot d\Sigma = 0$$

$$\text{and } \boldsymbol{u}(x, 0) = \boldsymbol{u}_0(x) \qquad \text{for } x \in \Omega.$$

Here $d\Sigma$ is the surface area vector. Note that if Ω is multiconnected then the divergence free condition would dictate that the sum of the flux through all the components of $\partial\Omega$ be zero. However, most of the existence theorems in the literature (for steady as well as unsteady flows) would require that the flux through each individual component of $\partial\Omega$ be seperately zero.

Let $(\boldsymbol{U}(x,t), P(x,t))$ be the basic solution field satisfying the governing equations and the boundary conditions. We are interested in studying the solution orbits nearby this given orbit in an appropriate function space. *A case of particular interest is when the basic solution orbit belongs to the global attractor* (See appendix B for a discussion on global attractors for the two dimensional case.). Let us introduce the change of variables $\boldsymbol{u} = \boldsymbol{U} + \boldsymbol{v}$ and $p = P + q$ in 2.1 to get,

$$\boldsymbol{v}_t + (\boldsymbol{U} \cdot \nabla)\boldsymbol{v} + (\boldsymbol{v} \cdot \nabla)\boldsymbol{U} + (\boldsymbol{v} \cdot \nabla)\boldsymbol{v} = -\nabla q + \nu\Delta\boldsymbol{v} \text{ in } \Omega \times (0,\infty)$$

$$\nabla \cdot \boldsymbol{v} = 0 \text{ in } \Omega \times (0,\infty) \tag{2.2}$$

$$\boldsymbol{v}(x,t) = 0 \text{ for } (x,t) \in \partial\Omega \times [0,\infty)$$

and

$$\boldsymbol{v}(x,0) = \boldsymbol{v}_0(x) \text{ for } x \in \Omega.$$

We will introduce the following function spaces:

$$j(\Omega) = \{\boldsymbol{u} : \Omega \to \boldsymbol{R}^n; \boldsymbol{u} \in \boldsymbol{C}_0^\infty(\Omega), \mathrm{div}\boldsymbol{u} = 0\}$$

$$\boldsymbol{H} = \{\boldsymbol{u} : \Omega \to \boldsymbol{R}^n; \boldsymbol{u} \in L^2(\Omega); \mathrm{div}\boldsymbol{u} = 0; \boldsymbol{u}(x) \cdot n = 0, \ x \in \partial\Omega\}$$

$$\boldsymbol{V} = \{\boldsymbol{u} : \Omega \to \boldsymbol{R}^n; \boldsymbol{u} \in \boldsymbol{H}^1(\Omega); \mathrm{div}\boldsymbol{u} = 0; \boldsymbol{u}(x) = 0, x \in \partial\Omega\}.$$

Here $\boldsymbol{H}^m(\Omega)$ is the Sobolev space of vectorfields whose distributional derivatives of order up to m are in $L^2(\Omega)$. It can be shown that for bounded domains the spaces $\boldsymbol{H}$ and $\boldsymbol{V}$ are respectively the closures of $j(\Omega)$ in the norms of $L^2(\Omega)$ and $\boldsymbol{H}^1(\Omega)$ [32, 53, 87]. One should note that due to the Poincare lemma (which applies when Ω is bounded in one direction) the norm of $\boldsymbol{H}^1(\Omega)$ and the norm obtained from the Dirichlet integral $\|\boldsymbol{u}\|^2 = \int_{\partial\Omega} \nabla\boldsymbol{u} \cdot \nabla\boldsymbol{u} \, dx$ are

12

equivalent in $\boldsymbol{V}$. The trace $\boldsymbol{u}(x) \cdot n, x \in \partial\Omega$ can be characterized as an element of $\boldsymbol{H}^{-\frac{1}{2}}(\partial\Omega)$ when $\boldsymbol{u}$ and $\mathrm{div}\boldsymbol{u}$ are square integrable [17].

Let us now define the Stokes operator. Consider the symmetric bilinear form $a(\cdot,\cdot)$ defined by

$$a(\boldsymbol{u},\boldsymbol{v}) = \nu \int_\Omega \nabla\boldsymbol{u} \cdot \nabla\boldsymbol{v}\,dx.$$

Then it is immediate that

$$|a(\boldsymbol{u},\boldsymbol{v})| \leq \nu\|\boldsymbol{u}\|_V\|\boldsymbol{v}\|_V, \qquad \forall \boldsymbol{u},\boldsymbol{v} \in \boldsymbol{V}$$

by Schwartz inequality and

$$a(\boldsymbol{u},\boldsymbol{u}) = \nu\|\boldsymbol{u}\|_V^2, \qquad \forall \boldsymbol{u} \in \boldsymbol{V}.$$

Hence by Lax-Milgram lemma there exists an isomorphism onto map $A \in \mathcal{L}(\boldsymbol{V};\boldsymbol{V}')$ such that

$$a(\boldsymbol{u},\boldsymbol{v}) = <A\boldsymbol{u},\boldsymbol{v}>_{V\times V'}, \qquad \forall \boldsymbol{u},\boldsymbol{v} \in \boldsymbol{V}$$

where $V' = \mathcal{L}(V;\boldsymbol{R})$ the dual of V.

We can define a restriction of this operator $\hat{A}$ in the following way. For a given $\boldsymbol{u} \in V$, if there exits $\boldsymbol{g} \in \boldsymbol{H}$ such that

$$a(\boldsymbol{u},\boldsymbol{v}) = (\boldsymbol{g},\boldsymbol{v})_H, \qquad \forall \boldsymbol{v} \in \boldsymbol{V}$$

then $\hat{A}\boldsymbol{u} = \boldsymbol{g}$ and $\boldsymbol{u} = D(\hat{A})$. The operator $\hat{A} \in \mathcal{L}(D(\hat{A});H)$ is again an isomorphism on to. The explicit form of $D(\hat{A})$ can be obtained from the Cattabriga regularity theorem [12, 79, 90]:

Theorem 2.1 *Let* $\Omega \subset \boldsymbol{R}^n, n = 2,3$ *be bounded and of class* $\boldsymbol{C}^r, r = max(s+2,2)$ *and let* $\boldsymbol{g} \in \boldsymbol{H}^s(\Omega), s \geq -1$ *be a given vector field. Then for the Stokes problem*

$$-\nu\Delta\boldsymbol{u} + \nabla q = \boldsymbol{g} \qquad in\ \Omega,$$

$$\nabla \cdot \boldsymbol{u} = 0 \qquad in\ \Omega,$$

$$and\ \boldsymbol{u}(x) = 0 \quad x \in \partial\Omega,$$

the unique solution $(u, q) \in [H^{s+2}(\Omega) \cap V] \times H^{s+1}(\Omega)/R$ and satisfies the estimate

$$\|u\|_{H^{s+2}(\Omega)} + \|q\|_{H^{s+1}(\Omega)/R} \leq C_0 \|g\|_{H^s(\Omega)}.$$

Thus defining $\hat{A}$ by $\hat{A}u = g$, we get,

$$\frac{1}{C_0}\|u\|_{H^2(\Omega)} \leq \|\hat{A}u\|_{L^2(\Omega)} \leq \|u\|_{H^2(\Omega)} \qquad \forall\, u \in H^2(\Omega) \cap V.$$

Hence we write $D(\hat{A}) = H^2(\Omega) \cap V$. The operator $A \in \mathcal{L}(V; V') \cap \mathcal{L}(D(A); H)$ is regularly accretive and selfadjoint[44].

Let us use the Riesz representation theorem to identify H with its dual H'. This gives us the continuous and dense embeddings:

$$D(A) \subset V \subset H \equiv H' \subset V' \subset D(A)'.$$

Rellich's lemma [1] leads to the compactness of these embeddings.

The fractional powers of the Stokes operator A^α, $\alpha \in R$ and $D(A^\alpha)$ can be defined easily using spectral resolution. We will denote by $D(A^{-\alpha})$ the dual of $D(A^\alpha)$. It follows from a theorem of Lions [56] that $D(A^{1/2}) = V$.

In order to treat the inertia terms we will introduce the following spaces. We define $E_m = H^m(\Omega) \cap H$ with $0 \leq m \leq r$. The Hodge decomposition theorem[11] leads to the existence of the orthogonal projector P_H from $L^2(\Omega)$ onto H. In fact P_H is linear continuous from $H^m(\Omega)$ onto $H^m(\Omega) \cap H = E_m$. Let us define

$$\mathcal{A}u = -\nu P_H \Delta u, \qquad \forall u \in E_{m+2} \cap V.$$

Then by the Cattabriga regularity theorem $\mathcal{A} : E_{m+2} \cap V :\to E_m$ is an isomorphism on to. Furthermore,

$$\mathcal{A}u = Au = -\nu P_H \Delta u, \qquad \forall u \in D(A) \text{ where } D(A) = E_2 \cap V.$$

The analysis of this monograph uses both real and complexification of the above and other function spaces.

2.1 Functional Framework For Unbounded Domains

There are delicate differences between the functional framework for bounded and unbounded domains. Moreover, for unbounded domains there is a further difference depending on whether the boundary of the flow domain is compact or noncompact. Fluid flow past obstacles are classified as exterior problems. In this case we require the boundary of the obstacle to be compact. In the cases of viscous flow through channels and pipes (which we call multichannel domains) the boundary is noncompact. In this section we will describe how the functional framework for these problems is connected with the unique solvability theorem for the Stokes problem. Let us now define a set of function spaces and then provide certain theorems for exterior as well as multichannel type domains. Let Ω be an *arbitrary* open set with class C^2 boundary $\partial\Omega$.

$$\boldsymbol{H}^1(\Omega) = \{\phi \in L^2(\Omega), \nabla\phi \in L^2(\Omega)\}$$

$$W_0(\Omega) = \text{ completion of } C_0^\infty(\Omega) \text{ vector fields in the norm } \|\nabla\phi\|_{L^2(\Omega)}$$

$$j(\Omega) = \{\phi \in C_0^\infty(\Omega); \nabla \cdot \phi = 0\}$$

$$\boldsymbol{H}(\Omega) = \text{ completion of } j(\Omega) \text{ in } L^2(\Omega)$$

$$\boldsymbol{H}^*(\Omega) = \{\phi \in L^2(\Omega); \nabla \cdot \phi = 0, \phi \cdot \boldsymbol{n}|_{\partial\Omega} = 0\}$$

$$\boldsymbol{V}_0(\Omega) = \text{ completion of } j(\Omega) \text{ in the norm of } \|\nabla\phi\|_{L^2(\Omega)}$$

$$\boldsymbol{V}_0^*(\Omega) = \{\phi \in W_0(\Omega); \nabla \cdot \phi = 0\}$$

$$\boldsymbol{V}_1(\Omega) = \text{ completion of } j(\Omega) \text{ in the norm of } \boldsymbol{H}^1(\Omega)$$

$$\boldsymbol{V}_1^*(\Omega) = \{\phi \in H_0^1(\Omega); \nabla \cdot \phi = 0\}$$

As noted in the previous section if Ω is a *bounded domain* then we have [53, 87, 32]

Theorem 2.2 *Let* $\Omega \subset \boldsymbol{R}^n, n = 2, 3$ *be a* **bounded** *open set with* C^2 *boundary* $\partial\Omega$. *Then*

(i) $\boldsymbol{H}(\Omega) = \boldsymbol{H}^*(\Omega)$

(ii) $\boldsymbol{V}_0(\Omega) = \boldsymbol{V}_0^*(\Omega)$

(iii) $\boldsymbol{V}_1(\Omega) = \boldsymbol{V}_1^*(\Omega)$

and the norms of $\boldsymbol{V}_0(\Omega)$ *and* $\boldsymbol{V}_1(\Omega)$ *are equivalent.*

However, as described below this result is not true for unbounded domains.

Let us first consider the **Exterior Problem**

Definition 2.1 *An arbitrary open set* $\Omega \subset \boldsymbol{R}^n, n = 2,3$ *is called an admissible exterior domain if*

(i) Ω *contains the complete neighborhood of infinity*

(ii) $\partial\Omega \in C^2$

(iii) $\boldsymbol{R}^n\backslash\Omega$ *is a compact domain.*

Theorem 2.3 *Let* Ω *be an admissible exterior domain. Then*

(i) $\boldsymbol{V}_0(\Omega) = \boldsymbol{V}_0^*(\Omega)$ *and*

(ii) $\boldsymbol{V}_1(\Omega) = \boldsymbol{V}_1^*(\Omega)$.

Theorem 2.4 *Let* Ω *be an admissible exterior domain. Then the problem of finding* $\boldsymbol{v} \in \boldsymbol{V}_1(\Omega) = \boldsymbol{V}_1^*(\Omega)$ *such that*

$$-\Delta\boldsymbol{v} + \nabla p = 0 \quad in \ \Omega$$

$$\nabla \cdot \boldsymbol{v} = 0 \quad in \ \Omega$$

$$\boldsymbol{v}|_{\partial\Omega} = 0, \quad \boldsymbol{v} \to 0 \ as \ |\boldsymbol{x}| \to 0$$

has the unique solution $\boldsymbol{v} = 0$.

These two theorems are proved in [32]. Note that the second theorem is the essential step in establishing the unique solvability result for the Stokes problem.

Let us now discuss an important aspect of the Stokes problem in the exterior domain. Let $\Omega \subset \boldsymbol{R}^n, n = 2,3$ be an admissible exterior domain. We consider the problem of finding $(\boldsymbol{u},p) : \Omega \to \boldsymbol{R}^n \times \boldsymbol{R}$ such that

$$-\Delta\boldsymbol{u} + \nabla p = 0 \quad in \ \Omega$$

$$\nabla \cdot \boldsymbol{u} = 0 \quad \text{in } \Omega \tag{2.3}$$

$$\boldsymbol{u}|_{\partial\Omega} = 0 \quad \text{and} \quad \boldsymbol{u} \to \boldsymbol{u}_\infty \text{ as } |\boldsymbol{x}| \to \infty$$

for a given $\boldsymbol{u}_\infty \neq 0$.

We then have the following well known paradox

Theorem 2.5 (Stokes Paradox)

There is no solution to (2.3), classical or generalized, in a two dimensional exterior domain $\Omega \subset \boldsymbol{R}^2$.

This result was conjectured by Stokes in 1851 [83]. Rigorous proofs for various classes of solutions were given in [72, 20, 13, 31]. We note here that a weak formulation of the problem gives the existence of a unique generalized solution $\boldsymbol{u}$ such that $\boldsymbol{u} - \boldsymbol{w}_\infty \in \boldsymbol{V}_0(\Omega)$ where $\boldsymbol{w}_\infty$ is some "constructed" solenoidal vectorfield which vanishes nearby $\partial\Omega$ and becomes equal to $\boldsymbol{u}_\infty$ beyond a certain finite distance from the obstacle (see [45] for such a construction.). However, in two dimensions, this information alone cannot force $\boldsymbol{u} - \boldsymbol{w}_\infty$ to decay to zero at infinity. This is due to the lack of an embedding theorem of the form $\boldsymbol{V}_0(\Omega) \subset L^p(\Omega)$ for any p. We have instead an embedding of the form

$$\int_\Omega \frac{|u(\boldsymbol{x}) - \boldsymbol{w}_\infty|^2}{|\boldsymbol{x}|^2 |\ln \boldsymbol{x}|^2} d\boldsymbol{x} \le C \int_\Omega |\nabla(\boldsymbol{u} - \boldsymbol{w}_\infty)|^2 d\boldsymbol{x}. \tag{2.4}$$

However this inequality will not force $\boldsymbol{u} - \boldsymbol{w}_\infty$ to decay at infinity since

$$\int_\Omega \frac{1}{|\boldsymbol{x}|^2 |\ln \boldsymbol{x}|^2} d\boldsymbol{x} < \infty.$$

The situation is different for the three dimensional exterior problem [45].

Theorem 2.6 *Let $\Omega \subset \boldsymbol{R}^3$ be an admissible exterior domain. Then $\exists$ a unique generalized solution $\boldsymbol{u}$ to 2.3 such that $\boldsymbol{u} - \boldsymbol{w}_\infty \in \boldsymbol{V}_0(\Omega)$.*

Note that if $\Omega \subset \boldsymbol{R}^3$ is an admissible exterior domain, then[45] $\boldsymbol{V}_0(\Omega) \subset L^6(\Omega)$. Moreover

$$\int_\Omega \frac{|u(\boldsymbol{x}) - \boldsymbol{w}_\infty|^2}{|\boldsymbol{x}|^2} d\boldsymbol{x} \le C \int_\Omega |\nabla(\boldsymbol{u} - \boldsymbol{w}_\infty)|^2 d\boldsymbol{x}. \tag{2.5}$$

For the three dimensional exterior domain we have

$$\int_\Omega \frac{1}{|x|^2} dx = \infty.$$

Hence, for this case $u(x) - w_\infty$ should decay to zero at infinity.

We thus see that the solvability problem for the exterior domain is connected to the decay properties of the solenoidal vector fields in $V_0(\Omega)$. For the two dimensional case the solenoidal vector fields in $V_0(\Omega)$ can even grow at infinity [31].

Although the behaviour of the solutions of the Stokes problem is undesirable in the above sense, they are some what well understood. However, the corresponding problem for the steady state Navier-Stokes equation in the two dimensional exterior domain remains as one of the profound open problems in Navier-Stokes theory (see section 3.1.1).

Let us now consider domains of the type shown in figure 2.1. These domains are defined so that outside a compact set, there are several outlets extending to infinity with specified divergence angles.

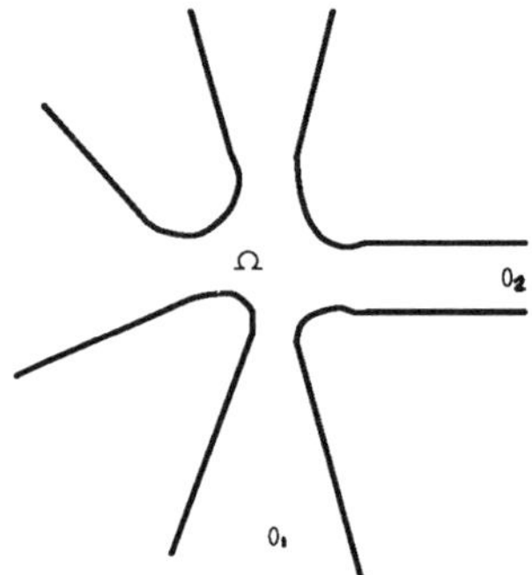

Figure2.1: Multichannel Domain

Definition 2.2 *A simply connected open set $\Omega \subset \mathbf{R}^2$ is called an admissible multichannel domain if it is the union of $N+1$ disjoint sets $\Omega_0 \cup \{\cup_{i=1}^N O_i\}, 2 \leq N < \infty$ defined in the following way:*

(i) Each $O_i, i = 1, \cdots, N$ is either a semi-infinite strip with width $d_i, 0 < d_i < \infty$ or a semi-infinite channel with constant divergent angle $\theta_i, 0 < \theta_i < \pi/2$. Here N is such that the

boundaries of two different channels do not intersect.

(ii) $\Omega_0 = \Omega \backslash \{\cup_{i=1}^{N} O_i\}$ and Ω_0 is bounded (not open).

(iii) All the components of the boundary $\partial\Omega$ are of class C^2.

It is possible to define a three dimensional multichannel domain in a similar manner. Let us now state a typical theorem showing how this class of problems are different from the interior as well as the exterior problems.

Theorem 2.7 *Let $\Omega \subset \mathbf{R}^2$ be an admissible multichannel domain with given N and positive outlet angles $0 < \theta_i < \pi/2$ for $i = 1, \cdots, N$. Then*

$$(i) \quad \boldsymbol{H}(\Omega) = \boldsymbol{H}^*(\Omega).$$

$$(ii) \quad \boldsymbol{V}_1(\Omega) = \boldsymbol{V}_1^*(\Omega).$$

$$(iii) \quad \boldsymbol{V}_0(\Omega) \subset \boldsymbol{V}_0^*(\Omega)$$

and

$$Dim \; \boldsymbol{V}_0^*(\Omega)/\boldsymbol{V}_0(\Omega) = N - 1.$$

This theorem is proved in [32] for $N = 2$ and in [50] for arbitrary N. It has been proven in these papers that for $\Omega \subset \mathbf{R}^3$ the results of (i) and (ii) of this theorem are replaced respectively by $\boldsymbol{H}(\Omega) \subset \boldsymbol{H}^*(\Omega)$ and $\boldsymbol{V}_1(\Omega) \subset \boldsymbol{V}_1^*(\Omega)$ with the codimension of the quotient spaces are again equal to $N - 1$.

This is a subtle mathematical result especially in the light of the Meyers-Serrin theorem [61] which states that if Ω is an arbitrary open set then the completion of $C_0^\infty(\Omega)$ vectorfields in a certain Sobolev norm coincides with the class of all vectorfields for which this norm is finite. The crucial difference here is that the divergence free condition is invoked before the completion for one class and after the completion for the other (denoted by the superscript $*$). It turns out that these two classes will not coincide for certain unbounded domains and the codimension between them will be equal to $N - 1$. It is interesting to see how such a geometrical aspect of the function spaces is related to the number of outlets and also to the number of additional conditions needed for the wellposedness of the Stokes problem. The next theorem indicates how we should formulate a uniquely solvable Stokes problem for such a domain.

Theorem 2.8 *Let $\Omega \subset \boldsymbol{R}^2$ be an admissible multichannel domain with given N and positive outlet angles $0 < \theta_i < \pi/2$ for $i = 1, \cdots, N$. Then the vector fields belong to $j(\Omega), \boldsymbol{H}(\Omega), \boldsymbol{V}_0(\Omega)$ and $\boldsymbol{V}_1(\Omega)$ do not carry flux through the outlets. That is if ϕ belongs to any of these classes, then*

$$\int_{\Sigma_i} \phi \cdot n dS_\Sigma = 0, \ \text{for } i = 1, \cdots, N$$

where Σ_i are cross sections (any curve connecting the lateral boundaries of O_i without being tangent at the boundaries) of the outlets O_i. Moreover, if $\phi \in \boldsymbol{V}_0^*(\Omega)$ with

$$\int_{\Sigma_i} \phi \cdot n dS_\Sigma \neq 0, \ \text{for some } i \text{ then } \phi \notin L^2(O_i).$$

From this theorem we deduce that a well posed Stokes problem in such an admissible two dimensional multichannel domain is to look for $v \in \boldsymbol{V}_0^*(\Omega)$ such that

$$-\Delta \boldsymbol{v} + \nabla p = 0 \ \text{ in } \Omega$$

$$\nabla \cdot \boldsymbol{v} = 0 \ \text{ in } \Omega$$

$$\boldsymbol{v}|_{\partial\Omega} = 0,$$

$$\boldsymbol{v} \to 0 \text{ as } |\boldsymbol{x}| \to 0 \text{ in each outlet } O_i$$

and

$$\int_{\Sigma_i} \boldsymbol{v} \cdot n dS_\Sigma = \alpha_i, \ \ i = 1, \cdots, N \text{ with } \sum_{i=1}^{N} \alpha_i = 0.$$

The flux rates α_i are the additional conditions which need to be prescribed. In three dimensional multichannel domains with divergent outlets, pressure tends to finite values at infinity. Hence in this case, instead of the flux conditions, we can prescribe the pressure drops to obtain a wellposed problem. Literature on this subject (which deals with two as well as three dimensional problems) is extensive and we refer the readers to [58, 59, 10, 42, 80, 81, 4]. Mathematical theory of time dependent flow through two dimensional multichannel domain is a new result and is developed in [34].

We will now develop a theory of Stokes operator suitable for arbitrary unbounded domains. We will denote by $a(\cdot,\cdot)$ the symmetric bilinear form,

$$a(\boldsymbol{u},\boldsymbol{v}) = \int_\Omega \nabla \boldsymbol{u} \cdot \nabla \boldsymbol{v}\,dx.$$

ν is not included for convenience. Let us define the Stokes operator in the folowing way.

Given $\boldsymbol{u} \in \boldsymbol{V}_1(\Omega)$, if there exists an element $\boldsymbol{g} \in \boldsymbol{H}(\Omega)$ such that

$$a(\boldsymbol{u},\boldsymbol{v}) = (\boldsymbol{g},\boldsymbol{v})_{L^2(\Omega)}, \forall \boldsymbol{v} \in \boldsymbol{V}_1(\Omega),$$

then we say that $\boldsymbol{u} \in D(A)$ and $A\boldsymbol{u} = \boldsymbol{g} \in R(A)$. We have in fact,

Theorem 2.9 *There exists a maximal accretive self adjoint operator $A \in \mathcal{L}(D(A); \boldsymbol{H}(\Omega))$ such that*

$$a(\boldsymbol{u},\boldsymbol{v}) = (A\boldsymbol{u},\boldsymbol{v})_{L^2(\Omega)}, \forall \boldsymbol{u} \in D(A), \forall \boldsymbol{v} \in \boldsymbol{V}_1(\Omega),$$

with $D(A) \subset \boldsymbol{V}_1(\Omega)$ dense. Moreover it is possible to extend this operator as

$$\hat{A} \in \mathcal{L}(\boldsymbol{V}_1(\Omega); \boldsymbol{V}_1(\Omega)')$$

such that

$$a(\boldsymbol{u},\boldsymbol{v}) = <\hat{A}\boldsymbol{u},\boldsymbol{v}>_{V_1(\Omega)' \times V_1(\Omega)}, \forall \boldsymbol{u},\boldsymbol{v} \in \boldsymbol{V}_1(\Omega).$$

We recall here that a densely defined linear accretive operator is maximal if and only if it is closed. Let us now prove the theorem.

Proof : We will obtain A as the strong limit of regularly accretive operators A_ϵ as $\epsilon \to 0$. Let us denote by $a_\epsilon(\cdot,\cdot)$ the symmetric bilinear form

$$a_\epsilon(\boldsymbol{u},\boldsymbol{v}) = \int_\Omega \nabla \boldsymbol{u} \cdot \nabla \boldsymbol{v}\,dx + \epsilon \int_\Omega \boldsymbol{u} \cdot \boldsymbol{v}\,dx, \text{ with } 0 < \epsilon < 1.$$

Using Schwartz inequality we obtain the continuity as

$$|a_\epsilon(\boldsymbol{u},\boldsymbol{v})| \le \|\boldsymbol{u}\|_{V_1(\Omega)}\|\boldsymbol{v}\|_{V_1(\Omega)}, \forall \boldsymbol{u},\boldsymbol{v} \in \boldsymbol{V}_1(\Omega), \text{ for } 0 \le \epsilon < 1.$$

The coerciveness $(\boldsymbol{V}_1(\Omega) - $ elliptic $)$ property is obtained as,

$$a_\epsilon(\boldsymbol{u},\boldsymbol{u}) \ge \epsilon\|\boldsymbol{u}\|^2_{V_1(\Omega)}, \forall \boldsymbol{u} \in \boldsymbol{V}_1(\Omega), \text{ for } 0 < \epsilon < 1.$$

Note that at $\epsilon = 0$, we have the following situation. The form is continuous,

$$|a(\boldsymbol{u}, \boldsymbol{v})| \leq \|\boldsymbol{u}\|_{V_1(\Omega)} \|\boldsymbol{v}\|_{V_1(\Omega)}, \forall \boldsymbol{u}, \boldsymbol{v} \in \boldsymbol{V}_1(\Omega).$$

But it is not $\boldsymbol{V}_1(\Omega) -$ elliptic . *This difficulty is due to the fact that the Poincare lemma does not hold for arbitrary unbounded domains.* At $\epsilon = 0$ the bilinear form satisfies

$$a(\boldsymbol{u}, \boldsymbol{u}) = \|\nabla \boldsymbol{u}\|^2_{L^2(\Omega)} = \|\boldsymbol{u}\|^2_{V_1(\Omega)} - \|\boldsymbol{u}\|^2_{L^2(\Omega)}, \forall \boldsymbol{u} \in \boldsymbol{V}_1(\Omega),$$

which is known as the Gårding inequality. Note that although $a(\cdot, \cdot)$ is not positive definite in $\boldsymbol{V}_1(\Omega)$, it is in fact positive

$$a(\boldsymbol{u}, \boldsymbol{u}) = \|\nabla \boldsymbol{u}\|^2_{L^2(\Omega)} \geq 0, \forall \boldsymbol{u} \in \boldsymbol{V}_1(\Omega).$$

Now, since the form $a_\epsilon(\cdot, \cdot)$ is symmetric, continuous and positive definite, we have the following standard results [85].

Proposition 2.1 : *There exists a self adjoint, regularly accretive onto map*
$A_\epsilon \in \mathcal{L}(D(A_\epsilon); \boldsymbol{H}(\Omega))$ *such that*

$$a_\epsilon(\boldsymbol{u}, \boldsymbol{v}) = (A_\epsilon \boldsymbol{u}, \boldsymbol{v})_{L^2(\Omega)}, \forall \boldsymbol{u} \in D(A_\epsilon), \forall \boldsymbol{v} \in \boldsymbol{V}_1(\Omega).$$

with $D(A_\epsilon) \subset \boldsymbol{V}_1(\Omega)$ dense. Moreover, it is posisible to extend this operator as an isomorphism on to map

$$\hat{A}_\epsilon \in \mathcal{L}(\boldsymbol{V}_1(\Omega); \boldsymbol{V}_1(\Omega)')$$

such that

$$a_\epsilon(\boldsymbol{u}, \boldsymbol{v}) = < \hat{A}_\epsilon \boldsymbol{u}, \boldsymbol{v} >_{V_1(\Omega)' \times V_1(\Omega)}, \forall \boldsymbol{u}, \boldsymbol{v} \in \boldsymbol{V}_1(\Omega).$$

We will now set,

$$A\boldsymbol{u} = A_\epsilon \boldsymbol{u} - \epsilon \boldsymbol{u}, \forall \boldsymbol{u} \in D(A) = D(A_\epsilon) \text{ and}$$

$$\hat{A}\boldsymbol{u} = \hat{A}_\epsilon \boldsymbol{u} - \epsilon \boldsymbol{u}, \forall \boldsymbol{u} \in \boldsymbol{V}_1(\Omega).$$

We can then easily show that the properties of A and $\hat{A}$ defined in the theorem hold. See Krein [44] or Tanabe [85] for a general treatment of a similar class of operators.

22

We note finally that since $a(\cdot,\cdot) : \boldsymbol{V}_0(\Omega) \times \boldsymbol{V}_0(\Omega) \to \boldsymbol{R}$ satisfies

$$|a(\boldsymbol{u},\boldsymbol{v})| \le \|\boldsymbol{u}\|_{V_0(\Omega)}\|\boldsymbol{v}\|_{V_0(\Omega)}, \forall \boldsymbol{u},\boldsymbol{v} \in \boldsymbol{V}_0(\Omega)$$

and

$$a(\boldsymbol{u},\boldsymbol{u}) = \|\boldsymbol{u}\|^2_{V_0(\Omega)}, \forall \boldsymbol{u} \in \boldsymbol{V}_0(\Omega),$$

we can define an isomorphism on to map

$$\breve{A} \in \mathcal{L}(\boldsymbol{V}_0(\Omega); \boldsymbol{V}_0(\Omega)')$$

such that

$$a(\boldsymbol{u},\boldsymbol{v}) = <\breve{A}\boldsymbol{u},\boldsymbol{v}>_{V_0(\Omega)' \times V_0(\Omega)}, \forall \boldsymbol{u},\boldsymbol{v} \in \boldsymbol{V}_0(\Omega).$$

The operators $A, \hat{A}$ and $\breve{A}$ are different characterizations of the Stokes operator and denoted simply as

$$A \in \mathcal{L}(D(A); \boldsymbol{H}(\Omega)) \cap \mathcal{L}(\boldsymbol{V}_1(\Omega); \boldsymbol{V}_1(\Omega)') \cap \mathcal{L}(\boldsymbol{V}_0(\Omega); \boldsymbol{V}_0(\Omega)').$$

Remark: Since $\boldsymbol{V}_1(\Omega) \subset \boldsymbol{V}_0(\Omega)$ we have $\boldsymbol{V}_0(\Omega)' \subset \boldsymbol{V}_1(\Omega)'$. However, since $\boldsymbol{V}_0(\Omega) \not\subset \boldsymbol{H}(\Omega)$ we have $\boldsymbol{H}(\Omega)' \not\subset \boldsymbol{V}_0(\Omega)'$ and care must be taken in the interpretation of the Stokes operator in the above form.

Let us now note certain properties of the range $R(A)$ and the domain $D(A)$ of the Stokes operator.

Lemma 2.1 $R(A) \subset \boldsymbol{H}(\Omega)$ *is dense.*

Proof: Let $\boldsymbol{g} \in \boldsymbol{H}(\Omega)$ be an arbitrary element. Since for any $0 < \epsilon < 1, A_\epsilon \in \mathcal{L}(D(A); \boldsymbol{H}(\Omega))$ is an on to map, $\exists \boldsymbol{u}_\epsilon \in D(A)$ such that

$$\boldsymbol{g} = A_\epsilon \boldsymbol{u}_\epsilon = (A + \epsilon)\boldsymbol{u}_\epsilon.$$

Now,

$$\|\boldsymbol{g} - A\boldsymbol{u}_\epsilon\|_{L^2(\Omega)} = \epsilon\|\boldsymbol{u}_\epsilon\|_{L^2(\Omega)}.$$

Hence $R(A) \subset \boldsymbol{H}(\Omega)$ is dense. Let us now prove a regularity theorem. This will give us an explicit representation of $D(A)$.

Theorem 2.10 : *Let $\Omega \subset \mathbf{R}^n, n = 2, 3$ be an arbitrary open set with boundary $\partial\Omega$ uniformly C^3. Then for a given $\mathbf{g} \in R(A) \subset \mathbf{H}(\Omega)$, the Stokes problem of finding $(\mathbf{u}, p) : \Omega \to \mathbf{R}^n \times \mathbf{R}$ such that*

$$-\Delta\mathbf{u} + \nabla p = \mathbf{g} \ in \ \Omega$$

$$\nabla \cdot \mathbf{u} = 0 \ in \ \Omega$$

$$\mathbf{v}|_{\partial\Omega} = 0 \ and \ \mathbf{v} \to 0 \ as \ |\mathbf{x}| \to \infty$$

has a unique solution $\mathbf{u}$ which satisfies,

$$\mathbf{u} \in D(A) = H^2(\Omega) \cap \mathbf{V}_1(\Omega).$$

Moreover, if we define the graph norm,

$$\|\mathbf{u}\|_{D(A)} = \|\mathbf{u}\|_{L^2(\Omega)} + \|A\mathbf{u}\|_{L^2(\Omega)}, \forall \mathbf{u} \in D(A)$$

then $\exists C > 0$ such that,

$$C\|\mathbf{u}\|_{H^2(\Omega)} \leq \|\mathbf{u}\|_{D(A)} \leq \|\mathbf{u}\|_{H^2(\Omega)}.$$

Proof:

Existence and uniqueness of $\mathbf{u} \in D(A) \subset \mathbf{V}_1(\Omega)$ has been established earlier. Let us analyse the regularity. For $\mathbf{g} \in R(A) \subset \mathbf{H}(\Omega)$, the following fundamental estimate holds [45, 33].

$$\|D^2\mathbf{u}\|_{L^2(\Omega)} \leq C_1(\|\mathbf{g}\|_{L^2(\Omega)} + \|\nabla\mathbf{u}\|_{L^2(\Omega)}), \tag{2.6}$$

with

$$\mathbf{g} = A\mathbf{u}.$$

Moreover, since $\mathbf{u} \in \mathbf{V}_1(\Omega)$ and solves,

$$a(\mathbf{u}, \mathbf{v}) = (\mathbf{g}, \mathbf{v}), \forall \mathbf{v} \in \mathbf{V}_1(\Omega),$$

we have setting $\mathbf{v} = \mathbf{u}$,

$$a(\mathbf{u}, \mathbf{u}) = \|\nabla\mathbf{u}\|_{L^2(\Omega)}^2 = (\mathbf{g}, \mathbf{u}) = (A\mathbf{u}, \mathbf{u})_{L^2(\Omega)}.$$

Hence, by Schwartz inequality,

$$\|\nabla u\|_{L^2(\Omega)} \leq \|u\|_{L^2(\Omega)}^{1/2}\|Au\|_{L^2(\Omega)}^{1/2}, \forall u \in D(A).$$

From the above estimate and (2.6) we can also get,

$$\|D^2 u\|_{L^2(\Omega)} \leq C_2(\|Au\|_{L^2(\Omega)} + \|u\|_{L^2(\Omega)}).$$

Hence,

$$C\|u\|_{H^2(\Omega)} \leq \|u\|_{D(A)}.$$

The reverse inequality is immediate since $Au = -\boldsymbol{P}_H\Delta u, \forall u \in D(A)$, where $\boldsymbol{P}_H : L^2(\Omega) \to \boldsymbol{H}(\Omega)$ is the orthogonal projection operator.

$$\clubsuit$$

Let us now define the fractional powers of A. This can be done by first characterizing the fractional powers of the operator A_ϵ which has bounded inverse in $\boldsymbol{H}(\Omega)$. Note that for $\lambda \in \boldsymbol{C}$ the sesquilinear form $\mathcal{A}_\lambda[\cdot, \cdot] : \boldsymbol{V}_1(\Omega) \times \boldsymbol{V}_1(\Omega) \to \boldsymbol{C}$ defined by

$$\mathcal{A}_\lambda[\boldsymbol{u}, \boldsymbol{v}] = a(\boldsymbol{u}, \boldsymbol{v}) + \epsilon(\boldsymbol{u}, \boldsymbol{v})_{H(\Omega)} + \lambda(\boldsymbol{u}, \boldsymbol{v})_{H(\Omega)} \tag{2.7}$$

satisfies

$$|\mathcal{A}_\lambda[\boldsymbol{u}, \boldsymbol{v}]| \leq \text{ const. } \|\boldsymbol{u}\|_{V_1(\Omega)}\|\boldsymbol{v}\|_{V_1(\Omega)}, \forall \boldsymbol{u}, \boldsymbol{v} \in \boldsymbol{V}_1(\Omega). \tag{2.8}$$

Moreover the real part of the sesquilinear form

$$\Re\mathcal{A}_\lambda[\boldsymbol{u}, \boldsymbol{u}] = \|\nabla u\|_{H(\Omega)}^2 + (\epsilon + \Re\lambda)\|u\|_{H(\Omega)}^2, \forall \boldsymbol{u} \in \boldsymbol{V}_1(\Omega).$$

Thus if $\Re\lambda \geq 0$ and $0 < \epsilon < 1$

$$\Re\mathcal{A}_\lambda[\boldsymbol{u}, \boldsymbol{u}] \geq \epsilon\|u\|_{V_1(\Omega)}^2, \forall \boldsymbol{u} \in \boldsymbol{V}_1(\Omega). \tag{2.9}$$

Hence as before we can define an isomorphism on to map

$$A_\epsilon + \lambda \in \mathcal{L}(\boldsymbol{V}_1(\Omega); \boldsymbol{V}_1(\Omega)') \cap \mathcal{L}(D(A); \boldsymbol{H}(\Omega))$$

such that

$$\mathcal{A}_\lambda[u, v] = < A_\epsilon u + \lambda u, v >_{V_1(\Omega)' \times V_1(\Omega)}, \quad \forall u, v \in V_1(\Omega).$$

Take $u \in D(A)$ and write

$$((A_\epsilon + \lambda)u, u)_{H(\Omega)} = \|\nabla u\|^2_{H(\Omega)} + (\epsilon + \lambda)\|u\|^2_{H(\Omega)}.$$

We thus get

$$\|(A_\epsilon + \lambda)u\|_{H(\Omega)} \geq (\epsilon + \Re\lambda)\|u\|_{H(\Omega)}.$$

Since $A_\epsilon + \lambda$ is on to for $\Re\lambda \geq 0$ we get for these values of λ,

$$\|(A_\epsilon + \lambda)^{-1}\|_{\mathcal{L}(H(\Omega);H(\Omega))} \leq (\epsilon + \Re\lambda)^{-1}. \tag{2.10}$$

This will allow us to define both negative and positive fractional powers of A_ϵ (For such results for operators with similar properties see [44]). Hence for $0 < \theta < \infty$ we have

$$A_\epsilon^{-\theta} = \frac{sin\pi\theta}{\pi} \int_0^\infty s^{-\theta} R(s; -A_\epsilon) ds.$$

Here $A_\epsilon^{-\theta} \in \mathcal{L}(H(\Omega); H(\Omega))$ form a strongly continuous semigroup. That is $A_\epsilon^{-\alpha} A_\epsilon^{-\beta} = A_\epsilon^{-(\alpha+\beta)}$ $\alpha, \beta \in R^+$ and

$$\forall u \in H(\Omega), \|A_\epsilon^{-\theta} u - u\|_{H(\Omega)} \to 0 \text{ as } \theta \to 0.$$

The positive powers of A_ϵ can be defined by

$$A_\epsilon^\theta = (A_\epsilon^{-\theta})^{-1}, \theta \in R^+.$$

We then have

Lemma 2.2 *For $\theta \in R^+$*

(i) A_ϵ^θ is a closed operator with domain $D(A_\epsilon^\theta)$ dense in $H(\Omega)$.

(ii) for $0 < \alpha < \beta$, we have $D(A_\epsilon^\beta) \subset D(A_\epsilon^\alpha)$.

Let us restrict ourselves to $0 < \alpha < 1$. We then have [44]

26

Theorem 2.11 *For $u \in D(A_\epsilon)$ we get*

$$\|A_\epsilon^\alpha u - A_\eta^\alpha u\|_{H(\Omega)} \leq C|\epsilon - \eta|^\alpha \|u\|_{H(\Omega)}$$

where C is independent of ϵ and η.

This theorem allows us to take $\epsilon \to 0$ and define the fractional powers of A. Let us write

$$A^\alpha u = \frac{sin\pi\alpha}{\pi} \int_0^\infty s^{\alpha-1} AR(s; -A)u\,ds, u \in D(A)$$

This formula actually defines the restriction of A^α to $D(A)$. We then have [44]

Theorem 2.12 *For $0 < \alpha < 1$ and $0 < \epsilon < 1$ we have $D(A_\epsilon^\alpha) = D(A^\alpha)$. Moreover*

$$\|A_\epsilon^\alpha u - A^\alpha u\|_{H(\Omega)} \leq C\epsilon^\alpha \|u\|_{H(\Omega)}, \forall u \in D(A^\alpha)$$

Since the range $R(A_\epsilon^\alpha) = \boldsymbol{H}(\Omega)$ we conclude from the above theorem that the range $R(A^\alpha)$ is dense in $\boldsymbol{H}(\Omega)$.

Note that since A is a self adjoint positive operator (means $(Au, u)_{H(\Omega)} \geq 0$) with dense domain its spectral resolution can be writen as the Stieltjes integral

$$A = \int_0^\infty \lambda dE(\lambda)$$

The spectral resolution operators (of the identity) $E(\lambda)$ satisfy the following properties

(i) $E(\lambda) = 0$ for $\lambda < 0$.

(ii) $E(\lambda) \to I$ strongly in $\boldsymbol{H}(\Omega)$ as $\lambda \to \infty$.

(iii) $E(\lambda)E(\mu) = E(\mu)E(\lambda) = E(\lambda)$ for $\lambda < \mu$.

(iv) $E(\lambda)$ is self adjoint.

(v) $E(\lambda)$ is strongly left continuous.

We can deduce that

$$A_\epsilon = \int_0^\infty (\lambda + \epsilon)dE(\lambda).$$

The fractional powers of A_ϵ can then be defined by the following alternative form

$$A_\epsilon^\alpha = \int_0^\infty (\lambda + \epsilon)^\alpha dE(\lambda).$$

Taking $\epsilon \to 0$ we get

$$A^\alpha = \int_0^\infty \lambda^\alpha dE(\lambda), \quad 0 < \alpha < 1. \tag{2.11}$$

We can also show that [44]

Theorem 2.13 $\boldsymbol{V}_1(\Omega) = D(A^{1/2})$ *and*

$$a(\boldsymbol{u}, \boldsymbol{v}) = (A^{1/2}\boldsymbol{u}, A^{1/2}\boldsymbol{v})_{H(\Omega)}, \quad \forall \boldsymbol{u}, \boldsymbol{v} \in D(A^{1/2}).$$

Remark:

For further properties of the Stokes operator in unbounded domains and in $L^p(\Omega)$ spaces see the works of Giga and his collegues[26, 27].

Chapter 3

The Linearized Operator And Its Spectral Properties

In this chapter we will develop a spectral theory for the linearized operator obtained by perturbing the steady state Navier-Stokes equations about a stationary basic field. We will establish a general spectral theorem for the linearized operator. The results for bounded domains presented in this chapter generalizes the previous works [69, 73] to larger classes of problems in a general functional frame work. Spectral theory for exterior problem will be discussed in section 3.3. For similar spectral theorems in other hydrodynamic problems see [30]. Analysis of the spectrum requires the knowledge of the existence and regularity (and for unbounded geometries the farfield decay) of the basic field. Hence we will first consider the problem of stationary hydrodynamics.

3.1 The Stationary Basic Field

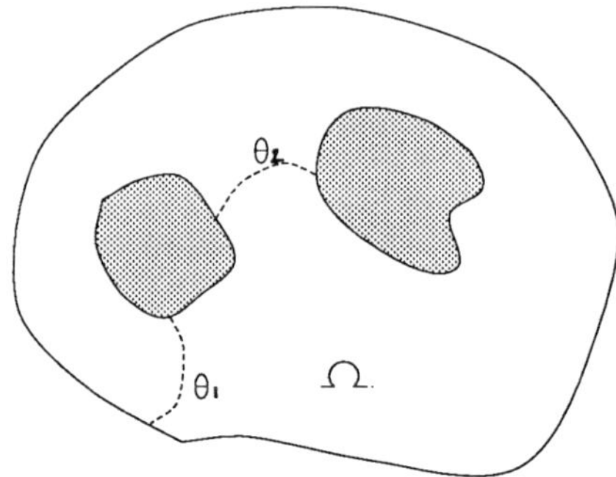

Figure 3.1: Multiconnected Domain

Let us develop a theory for the stationary hydrodynamics in a multiconnected bounded container. We will first consider certain important properties of the curl operator in such domains and the corresponding Hodge type decomposition of $L^2(\Omega)$ and higher order Sobolev spaces. These results are useful in constructing suitable extensions of the boundary data in to the interior of the container as solenoidal vectorfields with other needed properties.

Let Ω be an open bounded set of $\boldsymbol{R}^n, n = 2$ or 3. We assume that Ω be connected, of class $\boldsymbol{C}^r, r \geq m + 2, m \geq 0$ and its boundary $\partial\Omega$ has a finite number of connected components $\partial\Omega_1, \cdots, \partial\Omega_k, k \geq 1$. If the open set Ω is multiconnected we can make it simply connected by introducing a finite number of smooth cuts in the following way. Let $\Theta_1, \cdots, \Theta_N$ be $(n-1)$ dimensional manifolds of class $\boldsymbol{C}^r$ such that $\Theta_i \cap \Theta_j = \phi$ for $i \neq j$ and the open set $\Omega\backslash\Theta$, $\Theta = \bigcup_{i=1}^{N} \Theta_i$ is simply connected and Lipschtzian (see figure 3.1). Let us define[21],

$$\boldsymbol{H}_0 = \{v \in \boldsymbol{H}, \int_{\Theta_i} v \cdot d\Theta_i = 0, i = 1, \cdots, N\}.$$

Then $L^2(\Omega)$ can be decomposed as an orthogonal sum

$$L^2(\Omega) = \boldsymbol{H}_0 \oplus \ker(\mathrm{curl}).$$

Lemma 3.1 *For $m \geq 0$,*

$$curl\boldsymbol{H}^{m+1}(\Omega) = \{u \in \boldsymbol{H}^m(\Omega), div \cdot u = 0, \int_{\partial\Omega_i} u \cdot d\Sigma_i = 0, i = 1, \cdots, k\} \qquad (3.1)$$

and curl is an isomorphism from $\boldsymbol{H}^{m+1}(\Omega) \cap \boldsymbol{H}_0$ onto curl $\boldsymbol{H}^{m+1}(\Omega)$ which is equipped with the norm of $\boldsymbol{H}^m(\Omega)$.

Proof

We will use the following results [21, 17] :

Proposition 3.1 *curl $[\boldsymbol{H}^1(\Omega) \cap \boldsymbol{H}_0] = $ curl $\boldsymbol{H}^1(\Omega)$ and*

$$curl\boldsymbol{H}^1(\Omega) = \{\boldsymbol{u} \in \boldsymbol{L}^2(\Omega); div \cdot \boldsymbol{u} = 0; \int_{\partial\Omega_i} \boldsymbol{u} \cdot d\Sigma_i = 0, \forall i\}.$$

Proposition 3.2 *For Ω of class $\boldsymbol{C}^r, r \geq m+1$,*

$$\boldsymbol{H}^m(\Omega) = \{\boldsymbol{u} \in \boldsymbol{L}^2(\Omega), curl\boldsymbol{u} \in \boldsymbol{H}^{m-1}(\Omega), div \cdot \boldsymbol{u} \in \boldsymbol{H}^{m-1}(\Omega), \gamma_\nu \boldsymbol{u} \in \boldsymbol{H}^{m-\frac{1}{2}}(\partial\Omega)\}$$

and there exists a constant $C_2 = C_2(m, \Omega)$ such that $\forall \boldsymbol{u} \in \boldsymbol{H}^m(\Omega)$,

$$\|\boldsymbol{u}\|_{\boldsymbol{H}^m(\Omega)} \leq C_2\{\|\boldsymbol{u}\|_{\boldsymbol{L}^2(\Omega)} + \|curl\boldsymbol{u}\|_{\boldsymbol{H}^{m-1}(\Omega)}.$$

$$+\|div \cdot \boldsymbol{u}\|_{\boldsymbol{H}^{m-1}(\Omega)} + \|\gamma_\nu \boldsymbol{u}\|_{\boldsymbol{H}^{m-\frac{1}{2}}(\partial\Omega)}\}$$

In particular for $m = 1$,

$$\|\boldsymbol{u}\|_{\boldsymbol{H}^1(\Omega)} \leq C_0\|curl\boldsymbol{u}\|_{\boldsymbol{L}^2(\Omega)} \quad \forall \boldsymbol{u} \in \boldsymbol{H}^1(\Omega) \cap \boldsymbol{H}_0.$$

Here γ_ν is the normal trace operator defined as

$$\gamma_\nu \boldsymbol{u} = \boldsymbol{u} \cdot n|_{\partial\Omega} \text{ for } \boldsymbol{u} \in \boldsymbol{C}(\bar{\Omega}).$$

Let us now prove the lemma. We will set

$$Y = \{\boldsymbol{u} \in \boldsymbol{H}^m(\Omega), \text{ div} \cdot \boldsymbol{u} = 0, \int_{\partial\Omega_i} \boldsymbol{u} \cdot d\Sigma_i = 0, i = 1, \cdots, k\}.$$

Since $\boldsymbol{H}_0$ is orthogonal to the kernel of the curl operator, we get

$$\text{curl}[\boldsymbol{H}^{m+1}(\Omega) \cap \boldsymbol{H}_0] = \text{curl}\boldsymbol{H}^{m+1}(\Omega) \subset \text{ curl } \boldsymbol{H}^1(\Omega).$$

Hence if $u \in \text{curl}\, \boldsymbol{H}^{m+1}(\Omega)$ then there exists $\boldsymbol{v} \in \boldsymbol{H}^{m+1}(\Omega) \cap \boldsymbol{H}_0$ such that $\boldsymbol{u} = \text{curl}\, \boldsymbol{v}$. Moreover proposition 3.1 implies that $\boldsymbol{u}$ satisfies

$$\text{div} \cdot \boldsymbol{u} = 0, \int_{\partial \Omega_i} \boldsymbol{u} \cdot d\Sigma_i = 0, i = 1, \cdots, k.$$

Thus noting that for $\boldsymbol{v} \in \boldsymbol{H}^{m+1}(\Omega)$, $\text{curl}\, \boldsymbol{v} \in \boldsymbol{H}^m(\Omega)$ we get $\boldsymbol{u} \in Y$. That is curl $\boldsymbol{H}^{m+1}(\Omega) \subset Y$. On the other hand if $\boldsymbol{u} \in Y$ then since $Y \subset \text{curl}\, \boldsymbol{H}^1(\Omega)$ (due to proposition 3.1), there exists $\boldsymbol{v} \in \boldsymbol{H}^1(\Omega) \cap \boldsymbol{H}_0$ such that $\boldsymbol{u} = \text{curl}\, \boldsymbol{v}$. Now we apply proposition 3.2 to $\boldsymbol{v}$ to get,

$$\|\boldsymbol{v}\|_{H^{m+1}(\Omega)} \leq C_2[\|\boldsymbol{v}\|_{L^2(\Omega)} + \|\boldsymbol{u}\|_{H^m(\Omega)}].$$

Thus $\boldsymbol{v} \in \boldsymbol{H}^{m+1}(\Omega) \cap \boldsymbol{H}_0$ and $\boldsymbol{u} \in \text{curl}\, [\boldsymbol{H}^{m+1}(\Omega) \cap \boldsymbol{H}_0]$. We thus conclude that $Y = \text{curl}\, [\boldsymbol{H}^{m+1}(\Omega) \cap \boldsymbol{H}_0]$. Now note that the proposition 3.2 implies that $\forall \boldsymbol{v} \in \boldsymbol{H}^{m+1}(\Omega) \cap \boldsymbol{H}_0$, $\exists$ constants $C_3, C_4 > 0$ such that

$$C_3 \|curl v\|_{H^m(\Omega)} \leq \|\boldsymbol{v}\|_{H^{m+1}(\Omega)} \leq C_4 \|curl v\|_{H^m(\Omega)}.$$

Hence curl is isomorphic from $\boldsymbol{H}^{m+1}(\Omega) \cap \boldsymbol{H}_0$ on to curl $\boldsymbol{H}^{m+1}(\Omega)$.

$$\clubsuit$$

Existence of generalized solutions for the steady state Navier-Stokes equations can be proven [45] either using the Leray-Schauder principle or by applying Brouwer's fixed point principle to the Galerkin projection. To extend these ideas to inhomogeneous boundary value problems, one simply uses a change of variable technique so that the unknown vectorfield will again satisfy homogeneous boundary condition. In this method however, we should be able to estimate all the resultant additional terms without any restriction on the data. This is an important step and is usually achieved by the construction of the **Hopf-extension operator** Λ_δ [37, 45, 21] .

Theorem 3.1 *Under the hypothesis of the present section, $\forall \delta > 0$, $\exists$ a continuous linear map Λ_δ from*

$$\overset{\circ}{\boldsymbol{H}}{}^{m-\frac{1}{2}}(\partial \Omega) = \{\boldsymbol{u}_b \in \boldsymbol{H}^{m-\frac{1}{2}}(\partial \Omega), \int_{\partial \Omega_i} \boldsymbol{u}_b \cdot d\Sigma_i = 0, \forall i\}$$

into $\boldsymbol{H}^m(\Omega), m \geq 1$ such that

$$(i) \quad div \cdot \Lambda_\delta u_b = 0 \quad in \quad \Omega,$$

$$(ii) \quad \gamma_0 \Lambda_\delta u_b = u_b \quad on \ \partial\Omega$$

where γ_0 is the trace operator mapping

$$\boldsymbol{H}^m(\Omega) \ onto \ \overset{o}{\boldsymbol{H}}{}^{m-\frac{1}{2}}(\partial\Omega)$$

and

$$(iii) \quad |b(\boldsymbol{v}, \Lambda_\delta u_b, \boldsymbol{v})| \leq \delta \|u_b\|_{H^{m-\frac{1}{2}}(\partial\Omega)} \|\boldsymbol{v}\|_V^2, \forall u_b \in \overset{o}{\boldsymbol{H}}{}^{m-\frac{1}{2}}(\partial\Omega), \forall \boldsymbol{v} \in \boldsymbol{V}.$$

Proof :

Let us consider the problem of finding $(\psi, q) : \Omega \rightarrow R^n \times R$ such that,

$$-\Delta\psi + \mathrm{grad}\, q = 0 \quad in \quad \Omega$$

$$\mathrm{div}\,\psi = 0 \quad in \quad \Omega$$

$$\psi = u_b \ on \ \partial\Omega \ with \ u_b \in \overset{o}{\boldsymbol{H}}{}^{m-\frac{1}{2}}(\partial\Omega).$$

Then by Cattabriga regularity theorem there exists a unique $\psi \in \boldsymbol{H}^m(\Omega)$ that satisfies the boundary data. The lemma 3.1 will then imply that $\psi \in \mathrm{curl}\,[\boldsymbol{H}^{m+1}(\Omega) \cap \boldsymbol{H}_0]$. That is $\exists\ \boldsymbol{v} \in \boldsymbol{H}^{m+1}(\Omega) \cap \boldsymbol{H}_0$ such that $\psi = \mathrm{curl}\ \boldsymbol{v}$. The rest of the proof is achieved by the introduction of a smooth cut off function as in the references quoted above (see also the section 4.3 on the periodic basic field).

♣

The Hopf extension operator enables us to use the change of variables $\boldsymbol{u} = \boldsymbol{v} + \Lambda_\delta u_b$ to convert the inhomogeneous boundary value problem to a homogeneous one for the unknown $\boldsymbol{v}$ with forcing and certain additional terms. These additional terms can be sharply estimated without any condition on u_b due to the properties of the Hopf extension operator. We thus obtain an a priori estimate for the Dirichlet integral (enstrophy) of the unknown $\boldsymbol{v}$ with respect to data. The existence of solutions can then established using the Leray -Schauder theorem. The regularity can be proved by bootstrapping with the aid of the cattabriga regularity theorem. Further refinement is possible with regard to the generic solvability using the infinite dimensional version of the Sard's theorem. This particular result has been obtained by observing that the nonlinear map defined for the Leray-Shauder existence theorem is Fredholm. We thus state the

Theorem 3.2 *[45, 33, 21, 87] Let Ω be of class C^{m+1} and $u_b \in H^{m-\frac{1}{2}}(\partial\Omega), m \geq 1$ such that $\int_{\partial\Omega} u_b \cdot d\Sigma = 0$. If Ω is multiply connected such that $\partial\Omega = \bigcup_{i=1}^{k} \partial\Omega_i$ then we require that $\int_{\partial\Omega_i} u_b \cdot d\Sigma_i = 0, i = 1, \cdots, k$. Then there exists a solution $(U, P) \in H^m(\Omega) \times H^{m-1}(\Omega)/R$. For sufficiently small data this solution is unique. If $\partial\Omega$ is of class C^∞ and $u_b \in C^\infty(\partial\Omega)$ then $(U, P) \in C^\infty(\bar{\Omega}) \times C^\infty(\bar{\Omega})/R$. Moreover for any particular value of the Reynold's number, generically(for almost all u_b) there are only a finite number of stationary solutions.*

Remark:

In the case of multiply connected domains we have required that the flux through each connected component of the boundary to be individually zero. However the divergence free condition for the velocity field implies only the sum of these flux to be zero. Solvability theorem for the Navier-Stokes equations with this general flux condition seems to be open.

3.1.1 Exterior Hydrodynamics: Resolutions And Open Problems

In this section we will describe the existence theory of steady viscous flow past two and three dimensional obstacles. Let $\Omega \subset R^n, n = 2, 3$ be an admissible exterior domain. The problem is to find $(u, p) : \Omega \to R^n \times R$ such that

$$-\nu\Delta u + (u \cdot \nabla)u + \nabla p = 0 \text{ in } \Omega$$

$$\nabla \cdot u = 0 \text{ in } \Omega$$

$$u|_{\partial\Omega} = 0 \text{ and } u \to u_\infty \text{ as } |x| \to \infty,$$

with $\boldsymbol{u}_\infty = (1, 0, 0)$. We will begin by constructing the *Leray solutions* [52] for this problem. Let us construct a sequence of nested domains $\Omega_R = \Omega \cap \{|\boldsymbol{x}| < R\}$ as shown in figure 3.1.1. We consider the problem of finding $(\boldsymbol{u}_R, p_R) : \Omega_R \to \boldsymbol{R}^n \times \boldsymbol{R}$ with boundary conditions $\boldsymbol{u}_R|_{\partial\Omega} = 0$ and $\boldsymbol{u}_R|_{|\boldsymbol{x}|=R} = \boldsymbol{u}_\infty$. This is a problem in a bounded container and hence the results of the previous section apply. In fact for this problem Leray showed that

Theorem 3.3 $\forall \nu > 0$, $\exists$ *atleast one solution* $\boldsymbol{u}_R \in \boldsymbol{V}_0(\Omega_R)$ *such that*

$$\int_{\Omega_R} |\nabla \boldsymbol{u}_R|^2 dx \leq C(\nu). \tag{3.2}$$

Here $C(\nu)$ is independent of R.

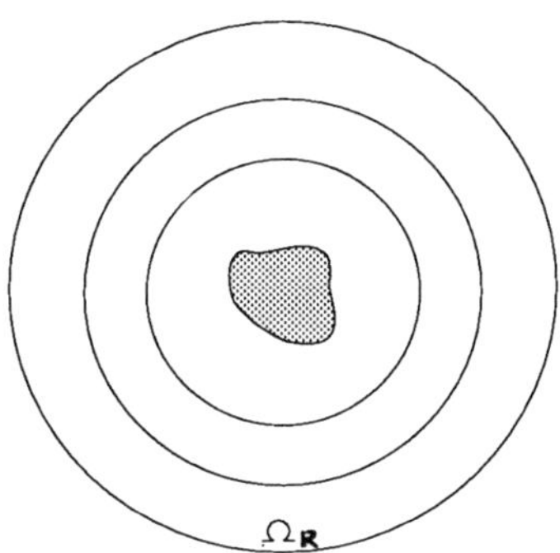

Figure 3.1.1: Leray Construction

The Leray solution $\boldsymbol{u}_L \in \boldsymbol{V}_0(\Omega)$ is then the weak limit of a suitable subsequence $\{\boldsymbol{u}_{R_i}\}$ in $\boldsymbol{V}_0(\Omega)$. Note that from the definition of the space $\boldsymbol{V}_0(\Omega)$, we have $\boldsymbol{u}_L|_{\partial\Omega} = 0$ and hence the boundary condition on the body is satisfied in the sense of trace. The important question then is whether the farfield condition $\boldsymbol{u}_L \to \boldsymbol{u}_\infty$ as $|\boldsymbol{x}| \to \infty$ is satisfied. We note here that although each of Leray's approximate solutions $\hat{\boldsymbol{u}}_{R_i}$ satisfies $\hat{\boldsymbol{u}}_{R_i}|_{|\boldsymbol{x}|=R_i} = \boldsymbol{u}_\infty$, since the limit is taken in the $\boldsymbol{V}_0(\Omega)$-space, the farfield condition may be lost in the limit. Moreover $\boldsymbol{u}_L$ may even be a trivial solution which ignores the farfield condition.

Let us first discuss the three dimensional problem for which these questions have been completely resolved [46, 19, 5]. As remarked in chapter 2 the estimate 3.2 should imply that,

$$\int_{\Omega_R} |\boldsymbol{u}_L - \boldsymbol{u}_\infty|^6 dx \leq C_1(\nu) \quad \text{and}$$

$$\int_{\Omega_R} \frac{|u_L - u_\infty|^2}{|x|^2} dx \leq C_2(\nu)$$

with C_1, C_2 independent of R. As shown in [46, 19] this implies that $u_L \to u_\infty$ as $|x| \to \infty$ for the three dimensional exterior problem. This and the decay properties obtained by Babenko [5] are summarized in the following theorem.

Theorem 3.4 *Let $\Omega \subset \mathbf{R}^3$ be an admissible exterior domain with C^2 boundary $\partial\Omega$. Then $\forall \nu > 0$ and for a given u_∞ with $u_\infty \neq 0$, $\exists u_L \in V_0(\Omega)$ such that $u_L|_{\partial\Omega} = 0$, $u_L \in C^1(\bar{\Omega}) \cap C^\infty(\Omega)$ and satisfies the following decay conditions*

$$|u_L(x) - u_\infty| \leq C_3(\nu)|x|^{-1}(|x| - u_\infty \cdot x + 1)^{-1}$$

and

$$|\frac{\partial}{\partial x^i} u_L(x)| \leq C_4(\nu)|x|^{-3/2}(|x| - u_\infty \cdot x + 1)^{-3/2}.$$

Moreover, the energy relationship

$$\mathcal{F} \cdot u_\infty = 2\nu \int_\Omega |def\, u_L|^2 dx$$

holds. Here $\mathcal{F}$ is the drag force.

Note that this solution exhibits the *parabolic wake* behind the obstacle as expected from physical intuition. To see the wake behavior we simply set $u_\infty \cdot x = |x|\cos\theta$. Then

$$|u_L - u_\infty| \leq C|x|^{-1} \quad \text{for } |\theta| \leq \pi|x|^{-1/2} \text{ and}$$

$$|u_L - u_\infty| \leq C|x|^{-(1+\sigma)} \quad \text{for } |\theta| \leq \pi|x|^{-(1-\sigma)/2}, \ 0 \leq \sigma \leq 1.$$

This indicates that in a paraboloidal region behind the obstacle the decay is slower. This is known as the wake region.

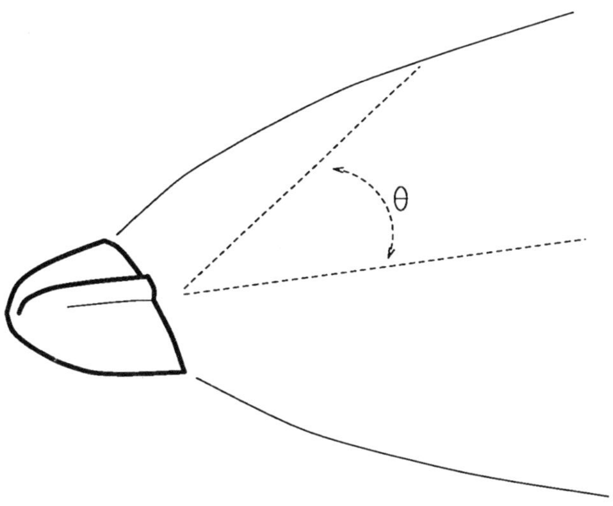

Figure 3.1.2: Leray-Finn-Babenko flow pattern

The Leray problem is far from resolved for the two dimensional exterior domains. As noted before the limit process will give us an element $u_L \in V_0(\Omega)$ which satisfies $u_L|_{\partial\Omega} = 0$. We note here that the Leray's approximate solution u_R^s for the Stokes problem in two dimensional exterior domain satisfies[3]

$$\int_{\Omega_R} |\nabla u_R^s|^2 dx \leq \frac{C|u_\infty|^2}{\ln R}, \quad R \geq 2$$

and hence in the limit we will get $u_L^s \equiv 0$ and the farfield boundary condition is completely ignored. Hence we have the following profound questions. *Is the Leray solution for the two dimensional steady state exterior problem nontrivial ?. If so, will it satisfy $u_L \to u_\infty$ as $|x| \to \infty$?* Partial answer to these questions is provided by the works of Gilbarg and Weinberger [28, 29] and of Amick [3]. Some of their key results are summarized in the theorem below.

Theorem 3.5 *Let (u_L, p_L) be the Leray solution for the two dimensional exterior problem. Then (Gilbarg and Weinberger [28, 29]):*

$$(i) \quad u_L \in L^\infty(\Omega).$$

(ii) $\exists p_0$ *such that* $p_L(r,\theta) \to p_0$ *as* $r = |\boldsymbol{x}| \to \infty$.

(iii) $\exists$ *a constant vector* $\hat{\boldsymbol{u}}_\infty \in \boldsymbol{R}^2$ *such that* $\boldsymbol{u}_L$ *converges to* $\hat{\boldsymbol{u}}_\infty$ *in the mean:*

$$\int_0^{2\pi} |\boldsymbol{u}_L(r,\theta) - \hat{\boldsymbol{u}}_\infty| d\theta \to 0 \ \ as \ r = |\boldsymbol{x}| \to \infty.$$

(iv) *If* $\hat{\boldsymbol{u}}_\infty = 0,$ *then* $\boldsymbol{u}_L \to 0$ *as* $|\boldsymbol{x}| \to \infty$.

Moreover, (Amick [3]):

(V) If $\boldsymbol{u}_L$ *is a symmetric solution (about the* $\boldsymbol{u}_\infty$- *axis) then* $\boldsymbol{u}_L \not\equiv 0$ *and as* $|\boldsymbol{x}| \to \infty$, $\boldsymbol{u}_L \to \hat{\boldsymbol{u}}_\infty$ *pointwise.*

We remark finally that some of the remaining major open problems now are

(a) Is the vector $\hat{\boldsymbol{u}}_\infty$ *equal to the given vector* $\boldsymbol{u}_\infty$ *?*

If so, then

(b) what is the farfield decay rate ?

3.2   Spectral Theory Of The Linearized Operator

Let us consider the spectral problem of finding $(\phi, q) : \Omega \to R^n \times R$ and $\lambda \in \boldsymbol{C}$ such that,

$$-\nu\Delta\phi + \nabla q + (\phi \cdot \nabla)\boldsymbol{U} + (\boldsymbol{U} \cdot \nabla)\phi = \lambda\phi \qquad in \quad \Omega$$

$$\nabla \cdot \phi = 0 \qquad in \quad \Omega \tag{3.3}$$

$$\phi(x) = 0, \qquad x \in \partial\Omega.$$

Here $\boldsymbol{U} \in \boldsymbol{H}^m(\Omega)$ is the basic field and Ω is a bounded domain with boundary $\partial\Omega$ is of class C^{m+2}. We will write this problem as an abstract spectral problem $\boldsymbol{L}\phi = \lambda\phi$, $\phi \in \boldsymbol{V}$ and establish the following *spectral theorem.*

Theorem 3.6 *The spectrum of* $\boldsymbol{L}$ *is discrete with finite multiplicity and can accumulate only at infinity. The corresponding eigenfunctions and generalized eigenfunctions are complete in* $D(A^s), s \in [0,1]$.

The proof of this theorem will be established in many steps.

We will first characterize $\boldsymbol{L}$ through a variational formulation of the problem (3.3). Let us define a sesquilinear form :

$$A_u[\phi,\psi] = a(\phi,\psi) + b(\phi,\boldsymbol{U},\psi) + b(\boldsymbol{U},\phi,\psi) \quad \text{where} \tag{3.4}$$

$$a(\phi,\psi) = \nu \int_\Omega \nabla\phi \cdot \nabla\bar{\psi}\,dx \quad \text{and}$$

$$b(\phi,\boldsymbol{\theta},\psi) = \int_\Omega (\phi \cdot \nabla\boldsymbol{\theta})\bar{\psi}\,dx.$$

Then from (3.3) we can derive the problem of finding $(\phi,\lambda) \in \boldsymbol{V} \times \boldsymbol{C}$ such that,

$$A_u[\phi,\psi] = \lambda(\phi,\psi)_{L^2(\Omega)}, \qquad \forall\psi \in \boldsymbol{V}.$$

Theorem 3.7 *Let $\boldsymbol{U} \in \boldsymbol{H}^3(\Omega)$, then the sesquilinear form $A_u[\cdot,\cdot] : \boldsymbol{V} \times \boldsymbol{V} \to \boldsymbol{C}$ is continuous*

$$|A_u[\phi,\psi]| \leq C_3\|\phi\|_V\|\psi\|_V, \quad \forall\phi,\psi \in \boldsymbol{V} \tag{3.5}$$

and satisfies the Gårding inequality

$$\Re A_u[\phi,\phi] \geq \nu\|\phi\|_V^2 - m_0\|\phi\|_{L^2(\Omega)}^2, \quad \forall\phi \in \boldsymbol{V} \tag{3.6}$$

for a suitable constant $m_0 > 0$.

Proof

Let us first note that the form $a(\cdot,\cdot) : \boldsymbol{V} \times \boldsymbol{V} \to \boldsymbol{C}$ satisfies

$$a(\phi,\phi) = \nu\|\phi\|_V^2, \qquad \forall\phi \in \boldsymbol{V}.$$

Also by Schwartz inequality we get the continuity:

$$|a(\phi,\psi)| \leq \nu\|\phi\|_V\|\psi\|_V \qquad \forall\phi,\psi \in \boldsymbol{V}.$$

Let us note certain properties of the form $b(\cdot,\cdot,\cdot) : \boldsymbol{V} \times \boldsymbol{V} \times \boldsymbol{V} \to \boldsymbol{C}$.

Lemma 3.2 $\forall \phi, \psi \in V$ *and* $U \in H^1(\Omega)$, *we have*

$$(i) \quad \Re b(U, \phi, \phi) = 0$$

$$(ii) \quad b(U, \phi, \psi) = -\overline{b(U, \psi, \phi)}$$

$$(iii) \quad |b(\phi, U, \psi)| \leq C_1 \|\phi\|_V \|U\|_{H^2(\Omega)} \|\psi\|_{L^2(\Omega)}$$

$$(iv) \quad |b(U, \phi, \psi)| \leq C_2 \|\phi\|_V \|U\|_{H^2(\Omega)} \|\psi\|_{L^2(\Omega)}$$

and if $U \in H^3(\Omega)$ *then*

$$(v) \quad \Re b(\phi, U, \phi) \geq -m_0 \|\phi\|_{L^2(\Omega)}^2.$$

Here m_0 is the greatest lower bound for the eigenvalues of the strain tensor

$$\frac{1}{2}\left(\frac{\partial}{\partial x^j} U_i + \frac{\partial}{\partial x^i} U_j\right)$$

of the basic field U.

Proof:

To prove (i) and (ii) we only need to establish the result for smooth vector fields since the form $b(\cdot, \cdot, \cdot)$ is continuous. Consider,

$$b(U, \phi, \psi) = \sum_{i,j=1}^{n} \int_\Omega U_j \frac{\partial \phi_i}{\partial x^j} \bar{\psi}_i dx$$

$$= \sum_{i,j=1}^{n} \int_\Omega \left\{\frac{\partial(U_j \phi_i \bar{\psi}_i)}{\partial x^j} - U_j \frac{\partial \bar{\psi}_i}{\partial x^j} \phi_i\right\} dx \quad \text{since div} \cdot U = 0.$$

Applying divergence theorem and noting that ϕ and ψ have zero trace we get

$$b(U, \phi, \psi) = -\overline{b(U, \psi, \phi)}$$

This implies also that

$$\Re b(U, \phi, \phi) = 0.$$

The form $b(\cdot, \cdot, \cdot)$ satisfies the following estimate[21]

$$|b(\theta, \phi, \psi)| \leq C\|\theta\|_{H^{s_1}(\Omega)} \|\phi\|_{H^{s_2+1}(\Omega)} \|\psi\|_{H^{s_3}(\Omega)} \tag{3.7}$$

where s_1, s_2 and s_3 are real nonnegative numbers such that,

$$s_1 + s_2 + s_3 \geq \frac{n}{2} \text{ if } s_i \neq \frac{n}{2}, i = 1, 2, 3$$

$$\text{and } s_1 + s_2 + s_3 > \frac{n}{2} \text{ if } s_i = \frac{n}{2} \text{ for some } i.$$

In our case $n = 2$ or 3 and hence to obtain (iii) we set $s_1 = 1, s_2 = 1, s_3 = 0$ and to obtain (iv) set $s_1 = 2, s_2 = 0, s_3 = 0$. The result (v) is proved in the monograph by Serrin[76]. The smoothness hypothesis required on U for his proof holds since $U \in H^3(\Omega)$ implies by Sobolev embedding theorem[1] that $U \in C^1(\Omega)$.

♣

The above results for the forms $a(\cdot, \cdot)$ and $b(\cdot, \cdot, \cdot)$ imply the required estimates for the sesquilinear form $A_u[\cdot, \cdot]$ in theorem 3.7

♣

Let us now define a new sesquilinear form $\hat{A}_u[\cdot, \cdot] : V \times V \to C$ as

$$\hat{A}_u[\phi, \psi] = A_u[\phi, \psi] + r(\phi, \psi)_{L^2(\Omega)} \quad \text{where } r > m_0.$$

We immediately get the coerciveness property

$$\Re \hat{A}_u[\phi, \phi] > \nu \|\phi\|_V^2, \quad \forall \phi \in V. \tag{3.8}$$

The continuity of this new sesquilinear form can be obtained as follows:

$$|\hat{A}_u[\phi, \psi]| \leq C_3 \|\phi\|_V \|\psi\|_V + r \|\phi\|_{L^2(\Omega)} \|\psi\|_{L^2(\Omega)}$$

by the Schwartz inequality. Now using Poincare's lemma we get

$$|\hat{A}_u[\phi, \psi]| \leq C_4 \|\phi\|_V \|\psi\|_V, \forall \phi, \psi \in V. \tag{3.9}$$

The continuity and the V-ellipticity established above enable us to use the Lax-Milgram lemma to associate $\bar{A}_u[\cdot, \cdot]$ with a duality pairing on $V' \times V$ as

$$\hat{A}_u[\phi, \psi] = < \hat{L}\phi, \psi >_{V' \times V}, \forall \psi, \phi \in V.$$

Here $\hat{L} : \boldsymbol{V} \to \boldsymbol{V}'$ is a continuous linear operator. If we write $\hat{L} = L + r$ then L is the abstract linear operator associated with the eigenvalue problem (3.3). Recall that the Stokes operator $A : \boldsymbol{V} \to \boldsymbol{V}'$ is associated with the coercive form $a(\cdot, \cdot)$ as

$$a(\boldsymbol{\phi}, \boldsymbol{\psi}) = < A\boldsymbol{\phi}, \boldsymbol{\psi} >_{V' \times V}, \quad \forall \boldsymbol{\phi}, \boldsymbol{\psi} \in V$$

and hence $L = A + L_u$ where L_u represents the first order terms

$$L_u \boldsymbol{\phi} = P_H \{ (\boldsymbol{U} \cdot \nabla) \boldsymbol{\phi} + (\boldsymbol{\phi} \cdot \nabla) \boldsymbol{U} \}, \quad \forall \boldsymbol{\phi} \in \boldsymbol{V}.$$

We have from equation (3.8)

$$\nu \|\boldsymbol{\phi}\|_V^2 < \Re \hat{A}_u[\boldsymbol{\phi}, \boldsymbol{\phi}] = \Re < \hat{L}\boldsymbol{\phi}, \boldsymbol{\phi} > \le \|\hat{L}\boldsymbol{\phi}\|_{V'} \|\boldsymbol{\phi}\|_V, \quad \forall \boldsymbol{\phi} \in \boldsymbol{V}.$$

That is $\|\hat{L}\boldsymbol{\phi}\|_{V'} > \nu \|\boldsymbol{\phi}\|_V$ and since the embedding $H \subset \boldsymbol{V}'$ is continuous, there exists a constant C_5 such that

$$C_5 \|\hat{L}\boldsymbol{\phi}\|_{L^2(\Omega)} \ge \|\hat{L}\boldsymbol{\phi}\|_{V'}, \forall \boldsymbol{\phi} \in D(A).$$

We thus get the injectivity of $\hat{L}$:

$$\|\hat{L}\boldsymbol{\phi}\|_{L^2(\Omega)} > \frac{\nu}{C_5} \|\boldsymbol{\phi}\|_V, \forall \boldsymbol{\phi} \in D(A).$$

Since the embedding of $\boldsymbol{V}$ in $\boldsymbol{H}$ is compact we deduce that $\hat{L}^{-1}$ is a compact operator in $\boldsymbol{H}$. Let us denote $R(\lambda; L) = (\lambda - L)^{-1}$ to be the resolvent of the operator L. Then

$$R(\lambda; L) = R(\lambda_0; L)[I + (\lambda_0 - \lambda)R(\lambda; L)] \tag{3.10}$$

for λ and λ_0 belonging to the resolvent set of L. If we set $\lambda_0 = -r$ and then $R(\lambda_0; L) = (-\hat{L})^{-1}$ and is compact. Thus $R(\lambda; L)$ is compact whenever it exists (since the operator inside the square bracket in equation (3.10) is bounded). Furthermore we have

$$[R(\lambda_0; L) - \mu]^k \boldsymbol{\psi} = (-\mu)^k R(\lambda_0; L)^k [(\lambda_0 - \frac{1}{\mu}) - L]^k \boldsymbol{\psi}. \tag{3.11}$$

Hence the generalized eigenfunction of the compact operator $R(\lambda_0; L)$ corresponding to the eigenvalue μ is the same as that for the operator L corresponding to the eigenvalue $\lambda_0 - 1/\mu$.

This enables us to conclude that

(i) L has discrete spectrum (since $R(\lambda_0; L)$ is compact and hence has discrete spectrum) which can accumulate only at infinity.

(ii) Completeness of the eigenfunctions (and generalized eigenfunctions) of L is established once we establish this result for the compact operator $R(\lambda_0; L)$.

In order to prove the completeness result we will derive certain regularity properties of the eigenfunctions and decay properties of the resolvent $R(\lambda; L)$

Lemma 3.3 *(i) If $U \in H^2(\Omega)$, then*

$$L_u \in \mathcal{L}(E_1; H) \cap \mathcal{L}(D(A^\alpha); D(A^{\alpha-1/2})), \alpha \in [0, 1/2].$$

(ii) If $U \in H^{k+1}(\Omega)$ with $k > 0$ then,

$$L_u \in \mathcal{L}(E_{k+1}; E_k).$$

Proof

Let $\psi \in H$ and write,

$$|(L_u\phi, \psi)_{L^2(\Omega)}| \le |b(U, \phi, \psi)| + |b(\phi, U, \psi)|,$$

and using the estimate (3.7)

$$\le (C_1 + C_2)\|U\|_{H^2(\Omega)}\|\phi\|_{H^1(\Omega)}\|\psi\|_{L^2(\Omega)}, \quad \forall \psi \in H.$$

Set $\psi = L_u\phi$ to get

$$\|L_u\phi\|_{L^2(\Omega)} \le (C_1 + C_2)\|U\|_{H^2(\Omega)}\|\phi\|_{H^1(\Omega)}.$$

This gives the continuity of the map $L_u : E_1 \to H$.

Let us now suppose $\psi \in V$ and write,

$$|< L_u\phi, \psi >_{V' \times V}| \le |b(U, \psi, \phi)| + |b(\phi, U, \psi)|$$

Again using the estimate 3.7 we get,

$$\le (C_{10} + C_{20})\|U\|_{H^2(\Omega)}\|\psi\|_{H^1(\Omega)}\|\phi\|_{L^2(\Omega)}, \quad \forall \psi \in V.$$

Hence noting that $\boldsymbol{V}' = D(A^{-1/2})$ we get,

$$\|L_u\phi\|_{D(A^{-1/2})} \le (C_1 + C_2)\|\boldsymbol{U}\|_{H^2(\Omega)}\|\phi\|_{L^2(\Omega)}.$$

Hence $L_u \in \mathcal{L}(L^2(\Omega); D(A^{-1/2}))$. Now the fact that

$$L_u \in \mathcal{L}(H; D(A^{-1/2})) \cap \mathcal{L}(D(A^{1/2}); H)$$

implies by interpolation theorem [54] $L_u \in \mathcal{L}(D(A^\alpha); D(A^{\alpha-1/2})), \alpha \in [0, 1/2]$.

Now for $k = 1$, we have $\phi, \boldsymbol{U} \in \boldsymbol{H}^2(\Omega)$ and then $(\boldsymbol{U}\cdot\nabla)\phi, (\phi\cdot\nabla)\boldsymbol{U} \in L^2(\Omega)$. Also

$$\frac{\partial}{\partial x^i}\{(\boldsymbol{U}\cdot\nabla)\phi\} = \frac{\partial \boldsymbol{U}}{\partial x^i}\cdot(\nabla\phi) + \boldsymbol{U}\cdot\frac{\partial}{\partial x^i}(\nabla\phi) \in L^2(\Omega), i = 1, \cdots, n.$$

This is true because by Sobolev embedding theorem the first derivatives of $\boldsymbol{U}$ and ϕ belong to $L^6(\Omega)$ and $\boldsymbol{U}, \phi$ are continuous in Ω. Similarly $\partial/\partial x^i\{(\phi\cdot\nabla)\boldsymbol{U}\} \in L^2(\Omega)$. The operator $P_H : \boldsymbol{H}^s(\Omega) \to \boldsymbol{E}_s, s \ge 0$ and therefore $L_u\phi \in \boldsymbol{E}_1$. If $k \ge 2$, then $\boldsymbol{H}^k(\Omega)$ is an algebra [1] and hence $\boldsymbol{U} \in \boldsymbol{H}^{k+1}(\Omega)$, $\nabla\phi \in \boldsymbol{H}^k(\Omega)$ implies that $(\boldsymbol{U}\cdot\nabla)\phi, (\phi\cdot\nabla)\boldsymbol{U} \in \boldsymbol{H}^k(\Omega)$. Hence $L_u\phi \in \boldsymbol{H}^k(\Omega) \cap \boldsymbol{H} = \boldsymbol{E}_k$.

♣

Theorem 3.8 *Let Ω be of class $C^r, r \ge m + 2, m \ge 0$ and the basic velocity field has the regularity $\boldsymbol{U} \in \boldsymbol{H}^m(\Omega)$. Then the eigenfunctions ϕ (and the generalized eigenfunctions) of L will have the regularity $\phi \in \boldsymbol{H}^m(\Omega) \cap \boldsymbol{V}$. Moreover if Ω is of class C^∞ and $\boldsymbol{U} \in C^\infty(\bar\Omega)$, then $\phi \in C^\infty(\bar\Omega)$.*

Proof

It was proved earlier that there exists $\phi \in \boldsymbol{V}$ for the eigenvalue problem $A\phi = \lambda\phi - L_u\phi$. The previous lemma implies that $\lambda\phi - L_u\phi \in \boldsymbol{H}$. Since A is an isomorphism from $\boldsymbol{E}_2 \cap \boldsymbol{V}$ onto $\boldsymbol{H}$ we can conclude that $\phi \in \boldsymbol{E}_2 \cap \boldsymbol{V}$. Again using the previous lemma we get $\lambda\phi - L_u\phi \in \boldsymbol{E}_1$ and hence the isomorphism of A would imply $\phi \in \boldsymbol{E}_3 \cap \boldsymbol{V}$. Proceeding in this manner we obtain $\phi \in \boldsymbol{E}_m \cap \boldsymbol{V}$.

Let us now consider the generalized eigenfunction ψ corresponding to λ that satisfies,

$$L\psi = \lambda\psi + \phi.$$

We again write

$$A\psi = -L_u\psi + \lambda\psi + \phi \in \boldsymbol{H}.$$

From this we deduce that $\psi \in D(A)$. Iterating this procedure we get $\psi \in \boldsymbol{E}_m \cap \boldsymbol{V}$. Regularity of the other generalized eigenfunctions can be successively established.

Now note that if the basic field is infinitely differentiable then we can carry out the above arguments to all m and obtain ϕ, ψ etc $\in \boldsymbol{E}_m \cap \boldsymbol{V}, \forall m$. From this we obtain the infinite differentiability using the Sobolev theorem.

♣

Theorem 3.9 *The resolvent of A satisfies the following estimates:*

$$(i) \quad \|R(\lambda; A)v\|_{D(A^s)} + \frac{|\lambda|}{|\lambda_I|}[2(|\lambda| - \lambda_R)]^{\frac{1}{2}}\|R(\lambda; A)v\|_{D(A^{s-\frac{1}{2}})}$$

$$+|\lambda|\|R(\lambda; A)v\|_{D(A^{s-1})} \leq \frac{|\lambda|}{|\lambda_I|}\|v\|_{D(A^{s-1})}, \quad s \in [0, 1]$$

for $\lambda_I \neq 0$.

$$(ii) \quad \|R(\lambda; A)\|_{\mathcal{L}(E_1; D(A))} \leq \frac{C}{|\lambda|^{\frac{1}{4}}}$$

for $\lambda_I \neq 0$ and $\lambda \neq 0$. Here λ_R and λ_I are respectively the real and imaginary parts of λ.

Proof

The properties of the operator A established earlier imply that its eigenfunctions $\{\psi_k\}_{k=1}^{\infty}$ are (orthonormal in $L^2(\Omega)$ and) complete in $D(A^{s/2})$. Therefore

$$R(\lambda; A)v = \sum_{k=1}^{\infty} \frac{(v, \psi_k)}{\lambda - \mu_k}\psi_k \quad \text{where} \quad \mu_k > 0 \tag{3.12}$$

are the eigenvalues of A and are real valued. Let $x \in [0, 2]$ and consider

$$A^{s-1+\frac{x}{2}} R(\lambda; A)v = \sum_{k=1}^{\infty} \frac{\mu_k^{\frac{x}{2}}}{(\lambda - \mu_k)}(v, \mu_k^{s-1}\psi_k)\psi_k.$$

Thus

$$\|R(\lambda; A)v\|^2_{D(A^{s-1+\frac{x}{2}})} = \sum_{k=1}^{\infty} \frac{\mu_k^x}{|\lambda - \mu_k|^2}|(A^{s-1}v, \psi_k)|^2.$$

Let

$$Y(x) = \sup_{k>0}\{\frac{\mu_k^x}{|\lambda - \mu_k|^2}\}$$

then

$$\|R(\lambda; A)v\|^2_{D(A^{s-1+\frac{x}{2}})} \le Y(x) \sum_{k=1}^{\infty} |(A^{s-1}v, \psi_k)|^2$$

$$= Y(x)\|v\|^2_{D(A^{s-1})}$$

We can easily verify that for $\lambda_I \neq 0$,

$$Y(0) = \frac{1}{|\lambda_I|^2}$$

$$Y(1) = \frac{1}{2(|\lambda| - \lambda_R)}$$

$$Y(2) = \frac{|\lambda|^2}{|\lambda_I|^2}.$$

This proves part (i). Part (ii) is proved in [39].

Remark

Note here that from the equation (3.12) the estimate

$$\|R(\lambda; A)\|_{\mathcal{L}(D(A^s);D(A^s))} \le \frac{1}{(\mu_1 - \lambda_R)} \tag{3.13}$$

can also be easily derived for $\lambda \notin (\mu_1, \infty)$.

Theorem 3.10 *Let Ω be of class C^3 and $\mathbf{U} \in \mathbf{H}^3(\Omega)$. Then the modified resolvent T_λ of the operator $R(-r; L)$ has the minimal growth property. That is:*

$$\|T_\lambda\|_{\mathcal{L}(D(A^s),D(A^s))} = O(\frac{1}{|\lambda|}), s \in [0, 1]$$

for large $|\lambda|$ along rays from the origin except along the negative real axis.

Proof

The modified resolvent of the operator $R(-r; L)$ is [2]

$$T_\lambda = [R(-r; L)^{-1} - \lambda]^{-1} = R(-r - \lambda; L)$$

and thus the minimal growth property of $R(\lambda; L)$ would imply the same for T_λ.

From the lemma 3.3 we have

$$L_u \in \mathcal{L}(D(A^\alpha); D(A^{\alpha - 1/2})) \cap \mathcal{L}(D(A); E_1)), \alpha \in [0, 1/2].$$

From the theorem 3.9 we have for large $|\lambda|$,

$$\|R(\lambda; A)\|_{\mathcal{L}(D(A^{\alpha - 1/2}); D(A^\alpha))} = \mathcal{O}(\frac{1}{|\lambda|^{1/2}}), \alpha \in [0, 1/2]$$

and

$$\|R(\lambda; A)\|_{\mathcal{L}(E_1; D(A))} = \mathcal{O}(\frac{1}{|\lambda|^{1/4}}).$$

We thus see that for large $|\lambda|$, $R(\lambda; A)L_u$ is a contraction in $\mathcal{L}(D(A^s); D(A^s))$. This implies that

$$[I + R(\lambda; A)L_u]^{-1} \in \mathcal{L}(D(A^s); D(A^s)).$$

Now for large $|\lambda|$ the estimate (3.13) gives,

$$\|R(\lambda; A)\|_{\mathcal{L}(D(A^s); D(A^s))} = \mathcal{O}(\frac{1}{|\lambda|})$$

except along the positive real axis. Hence noting that

$$R(\lambda; L) = [I + R(\lambda; A)L_u]^{-1}R(\lambda; A),$$

we obtain the minimal growth property of $R(\lambda; L)$.

♣

Lemma 3.4 *For large r the operator $R(-r; L) \in \mathcal{L}(D(A^s), D(A^s)), s \in [0, 1]$ is of finite double norm (Hilbert Schmidt class).*

Proof

We will note that the resolvent of A at $\lambda = -r$ is of finite double norm in $D(A^s)$. Let $\{\psi_k\}_{k=1}^{\infty}$ be the eigenfunctions of A orthonormal in $\mathbf{H}$. Then $\{\hat{\psi}_k\}_{k=1}^{\infty}$ with $\hat{\psi}_k/\mu_k^s$ is an orthonormal basis for $D(A^s)$. Hence $\|R(-r;A)\hat{\psi}_k\|_{D(A^s)} = (\mu_k + r)^{-1}$ and thus the double norm [2] of $R(-r;A)$ in $D(A^s)$ is

$$\||R(-r;A)|\| = \{\sum_{k=1}^{\infty} \frac{1}{(\mu_k + r)^2}\}^{\frac{1}{2}} < \infty.$$

This sum is finite since μ_k is of finite multiplicity. We observed earlier that for large r,

$$[I - R(-r;A)L_u]^{-1} \in \mathcal{L}(D(A^s), D(A^s)).$$

It is well known that the product composition of a bounded operator with an operator of finite double norm is also an operator of finite double norm. We thus use the form $R(-r;L) = [I - R(-r;A)L_u]^{-1}R(-r;A)$ to conclude that $R(-r;L)$ is of finite double norm in $D(A^s)$.

♣

Using the above properties of the resolvent $R(-r;L)$ (theorem 3.10 and lemma 3.4) and an estimate on the modified resolvents of such operators due to Carleman in combination with the Phragmen-Lindelof theorem, we can deduce from lemma 16.3 and theorem 16.4 of ref [2] that the span of the eigenfunctions (and generalized eigenfunctions) of $R(-r;L)$ (hence that of L) is complete in $D(A^s), s \in [0,1]$.

♣

3.3 Spectral Theory Of Three Dimensional Exterior Problem

In this section we will briefly describe a spectral theory due to Babenko[7] for the three dimensional exterior problem. This result was later used by him to study Karman vortex shedding[6] (see section 4.3.1).

Let us now formulate this spectral problem. Let $\Omega \subset \boldsymbol{R}^3$ be an admissible exterior domain. We will look for $(\phi, p) : \Omega \to \boldsymbol{R}^3 \times \boldsymbol{R}$ such that

$$-\frac{1}{\boldsymbol{R}_e}\Delta\phi + \nabla p + (\boldsymbol{u}_L \cdot \nabla)\phi + (\phi \cdot \nabla)\boldsymbol{u}_L = \lambda\phi \text{ in } \Omega$$

$$\nabla \cdot \phi = 0$$

$$\phi|_{\partial\Omega} = 0 \text{ and } \phi \to 0 \text{ as } |\boldsymbol{x}| \to \infty$$

Here $\boldsymbol{u}_L \in \boldsymbol{V}_0(\Omega)$ is the Leray-Finn-Babenko steady state solution to the Navier-Stokes equation at an arbitrary Reynold's number $\boldsymbol{R}_e$. The decay characteristics of $\boldsymbol{u}_L$ is given in theorem 3.4. As before we will denote this spectral problem as $\boldsymbol{L}\phi = \lambda\phi$. Let us define a set $\Lambda \subset \boldsymbol{C}$ as

$$\Lambda = \{\lambda; \lambda = \sigma + i\tau \text{ with } \tau^2 - \boldsymbol{R}_e\sigma \leq 0\}$$

Note that Λ here is a *region* bounded by a parabola and the origin $0 \in \Lambda$. We then have

Theorem 3.11 *For each Reynolds number $\boldsymbol{R}_e > 0$ we have*

$$Spectrum\ \boldsymbol{L} = \Lambda \cup \{\lambda_i\}$$

where the component $\{\lambda_i\}$ if nonempty will consists of a discrete set of finite multiplicity with accumulation possible on $\partial\Lambda$ or at infinity. Each λ_i will correspond to an eigenfunction $\phi_i \in H^2(\Omega) \cap \boldsymbol{V}_1(\Omega)$.

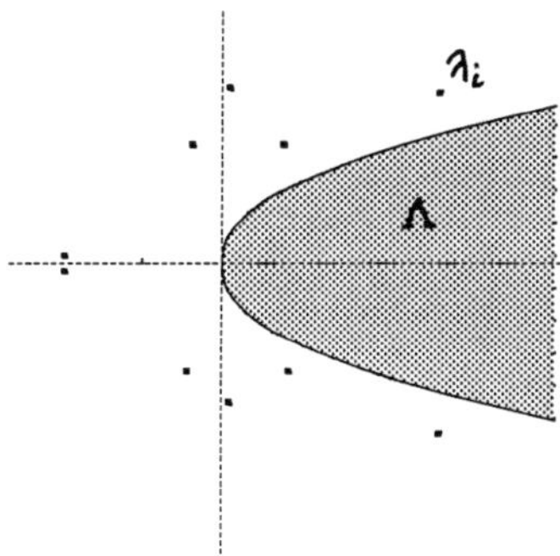

Figure 3.3: Babenko Spectrum

We note here that for this and other types of spectral problems associated with the linearized operator the eigenvalues appear as *complex conjugates* since the basic velocity about which we linearize the Navier Stokes equations is real. A comparison with the results for the bounded domain shows the additional nontrivial features of the exterior problem such as the existence of the *essential spectrum* Λ and that $0 \in \Lambda$ for all Reynolds numbers. Babenko states the following theorem regarding the Critical Reynold's number.

Theorem 3.12 *The eigenvalues $\lambda_i(\boldsymbol{R}_e)$ depend analytically on the Reynold's number. Moreover $\exists$ a Critical Reynold's number $0 < \boldsymbol{R}_{ec} < \infty$ such that for $\boldsymbol{R}_e < \boldsymbol{R}_{ec}$, the discrete part of the spectrum $\{\lambda_i\}$ can lie only in the right half plane $\Re\lambda > 0$.*

It is evident that this result is connected with the temporal stability of the Leray-Finn-Babenko steady solution.

Chapter 4

The Monodromy Operator And Its Properties

In this chapter we will analyze the properties of the semigroup associated with the evolution problems obtained by linearizing the Navier-Stokes equations about stationary and time dependent basic fields. The fundamental step in this analysis is the generation of the linear semigroup using the Stokes operator. From the properties of the Stokes operator (for bounded and unbounded domains) established in chapter 2, we can deduce that this operator generates a holomorphic semigroup. For bounded domains, the Stokes operator is a contraction semigroup with exponential time decay. This is not true for the case of unbounded domains and the decay theory is not completely resolved. Moreover exponential time decay can only be expected for domains which are bounded in one direction. For a detailed analysis of the Stokes semigroup in unbounded domains and in general spaces ($L^p(\Omega)$ and $C^{0+\alpha}(\Omega)$) see the works of Bemelmans [9] and Giga and his colleagues[25, 26, 27]. In the next section we will provide a basic theory of the Stokes semigroup in bounded domains. In section 4.2 we will consider the semigroup generated by the linearized operator $-L$. Section 4.3 provides a global existence theorem for the time periodic solutions to the Navier-Stokes equations. In section 4.4 we study the time evolution problem obtained by linearizing the Navier-Stokes equations about a time dependent basic solution. We then specialize our analysis to the case where the basic field is T-periodic in time. The time -T solution map of this linear evolution system is called the monodromy operator. It has several nice properties. The spectral properties of this operator is also studied. The spectral projection of the monodromy operator actually characterizes the tangent spaces of the stable and unstable manifolds and this result will be utilized later in chapters 6 and 7.

4.1   The Stokes Semigroup

Let us generate the Stokes semigroup (generated by $-A$) using the Hille-Yoshida theorem and then show that it can be extended as a holomorphic semigroup in a sector containing the positive real axis. The behavior of this semigroup at the origin with respect to various operator norms will also be established. Such estimates are needed in order to resolve the nonlinear hydrodynamic semigroup. Let us first obtain an estimate for the resolvent $R(\lambda; -A) = (\lambda + A)^{-1}$.

Lemma 4.1 *Let $\Omega \in R^n, n = 2$ or 3 be a bounded domain with boundary $\partial\Omega$ be of class C^2. Then the resolvent operator $R(\lambda; -A)$ satisfies*

$$\|R(\lambda; -A)\|_{\mathcal{L}(D(A^\alpha); D(A^\alpha))} \leq \frac{C_0}{|\lambda + \sigma|} \ for \ \lambda \in \Sigma_\delta \ with$$

$$\Sigma_\delta = \{\lambda \in C; |\arg(\lambda + \sigma)| \leq \frac{\pi}{2} + \delta\} \ and \ \alpha \in [-1, 1].$$

Here $0 < \sigma < \mu_1$ where μ_1 is the smallest eigenvalue of A and $\delta > 0$ is a suitable constant.

Proof

We have from the resolvent estimate obtained earlier in theorem 3.9 that,

$$\|R(\lambda; -A)\|_{\mathcal{L}(D(A^\alpha); D(A^\alpha))} < \frac{1}{\lambda_I} \ for \ \lambda_I \neq 0.$$

We can also easily obtain,

$$\|R(\lambda; -A)\|_{\mathcal{L}(D(A^\alpha); D(A^\alpha))} < \frac{1}{\lambda_R + \mu_1} \ for \ \lambda_R > -\mu_1.$$

Combining these two results we get

$$\|R(\lambda; -A)\|_{\mathcal{L}(D(A^\alpha);D(A^\alpha))} < \frac{2^{1/2}}{|\lambda + \sigma|} \text{ for } \lambda_R > -\sigma.$$

Now using the fact that $R(\lambda; -A)$ is a holomorphic operator we obtain by analytic continuation

$$\|R(\lambda; -A)\|_{\mathcal{L}(D(A^\alpha);D(A^\alpha))} \leq \frac{C_0}{|\lambda + \sigma|} \text{ for } \lambda \in \Sigma_\delta$$

Here Σ_δ is determined by the convergence of the Taylor expansions associated with this continuation (see figure 4.1).

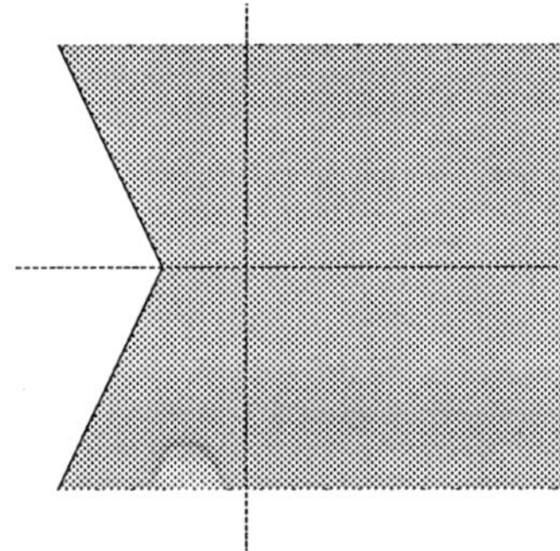

Figure 4.1: The Region Σ_δ

Theorem 4.1 *The operator $-A$ generates a strongly continuous contraction semigroup $S(t)$ in $D(A^\alpha)$ which can be extended as a holomorphic semigroup in an acute sector containing the positive real axis. Moreover,*

$$(i) \quad \|S(t)\|_{\mathcal{L}(D(A^\alpha),D(A^\alpha))} < e^{-\sigma t}, \quad t \geq 0, \text{ for } \alpha \in [-1,1]$$

$$(ii) \quad \|S(t)\|_{(B_1,B_2)} \leq Ct^{-\alpha}e^{-\sigma t}, \quad t > 0,$$

where the constant α is given as follows:

$$(a) \quad \text{If } B_1 = D(A^{\alpha_1}), B_2 = D(A^{\alpha_2}) \text{ with}$$

$$\alpha_1, \alpha_2 \in [-1,1] \text{ then } \alpha = \alpha_2 - \alpha_1 \geq 0$$

53

or

$$(b) \quad \text{If } B_1 = E_1, B_2 = D(A) \text{ then } \alpha = \frac{3}{4}.$$

Proof

When $\alpha_1 = \alpha_2 = \alpha \in [-1, 1]$, we have the estimate for the resolvent,

$$\|R(\lambda; -A)\|_{\mathcal{L}(D(A^\alpha), D(A^\alpha))} < \frac{1}{\lambda_R + \sigma} \text{ for } \lambda_R \geq -\sigma, \quad \sigma < \mu_1.$$

Hence by Hille-Yoshida theorem, $-A$ generates a C^0 semigroup with a decay rate given by (i). From the estimate on the resovent in lemma 4.1 we deduce that $S(t)$ can be extended as an analytic semigroup in a sector containing the positive real axis [67] (see figure 4.2).

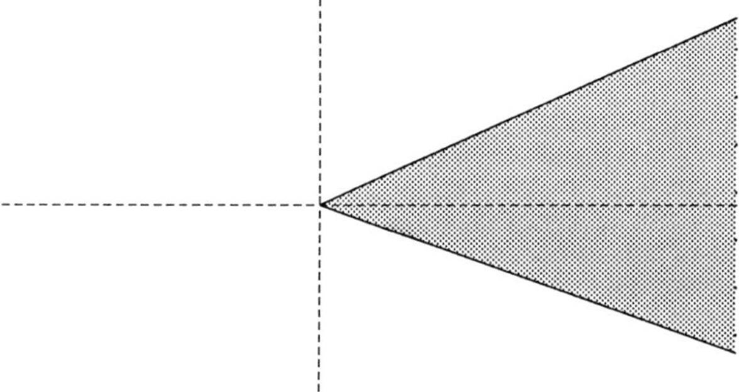

Figure 4.2: The Analyticity Domain For $S(t)$

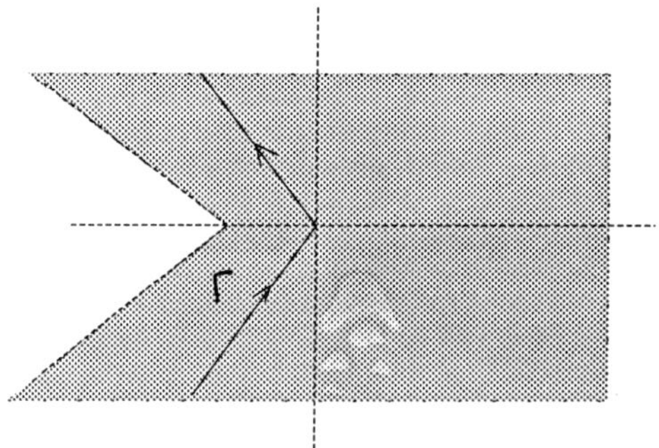

Figure 4.3: The Contour Γ

Moreover,

$$S(t)u = \frac{1}{2\pi i} \int_\Gamma e^{\lambda t} R(\lambda; -A)u \, d\lambda$$

with Γ consisting of two rays $xe^{i\theta}$ and $xe^{-i\theta}, 0 < x < \infty$ and $\frac{\pi}{2} < \theta < \pi$. Γ oriented so that λ_I increases along Γ (see figure 4.3).

The resolvent estimate of theorem 3.9 in the previous chapter then leads to

$$\|A^\alpha S(t)\|_{\mathcal{L}(D(A^{\alpha_1}),D(A^{\alpha_1}))} = \|S(t)\|_{\mathcal{L}(D(A^{\alpha_1}),D(A^{\alpha_2}))} \leq Ct^{-\alpha}e^{-\sigma t}, \text{ for } t > 0$$

and $\alpha = \alpha_2 - \alpha_1 \geq 0$.

In order to prove the result (ii)(b), we use the estimate (ii) of theorem 3.9. To estimate the semigroup in this case we can take the contour (see figure 4.4) as $\Gamma = \Gamma_1 \cup \Gamma_2 \cup \Gamma_3$ where

$$\Gamma_1 = \{xe^{i\theta}; t^{-1} \leq x \leq \infty\}$$

$$\Gamma_2 = \{t^{-1}e^{i\phi}; -\theta \leq \phi \leq \theta\} \text{ and}$$

$$\Gamma_3 = \{xe^{i\theta}; t^{-1} \leq x \leq \infty\} \text{ with } \frac{\pi}{2} < \theta < \pi$$

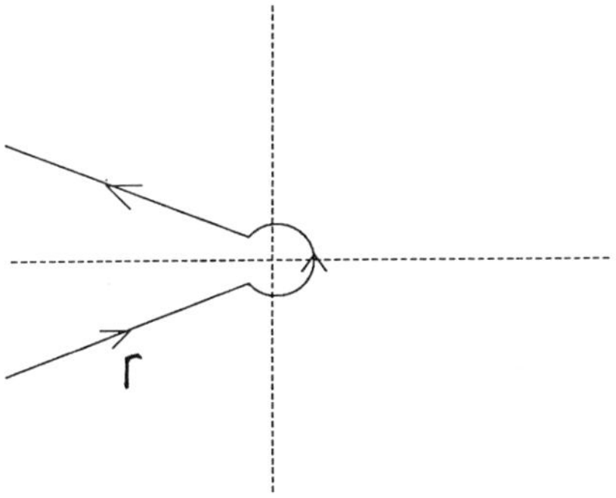

Figure 4.4: The Contour Γ

Remark: The analyticity of the semigroup $S(t)$ implies continuity in the uniform operator topology of $\mathcal{L}(D(A^{\alpha_1}); D(A^{\alpha_2})) \cap \mathcal{L}(E_1; D(A))$ for $t > 0$. See lemma 4.2 below for an example of such continuity estimate. These properties will be used later in establishing similar continuity results for the evolution operator.

Lemma 4.2 *For $\alpha = 1/2$ or 1, the Stokes semigroup $S(t)$ satisfies estimates of the type*

$$(i) \|AS(t)\|_{\mathcal{L}(E_{2\alpha-1};D(A^\alpha))} \le C_0 t^{-\alpha/2 - 5/4}, t > 0, \ and$$

$$(ii) \|S(t_2) - S(t_1)\|_{\mathcal{L}(E_{2\alpha-1};D(A^\alpha))} \le C_1(t_1^{-\alpha/2-1/4} - t_2^{-\alpha/2-1/4}), \quad 0 < t_1 \le t_2.$$

Proof

Due to the analyticity of $S(t)$ we can write,

$$AS(t) = \frac{1}{2\pi i} \int_\Gamma \lambda e^{\lambda t} R(\lambda; -A) d\lambda.$$

Hence

$$\|AS(t)\|_{\mathcal{L}(E_{2\alpha-1};D(A^\alpha))} \le \frac{1}{2\pi} \int_\Gamma |\lambda| e^{t\Re\lambda} \|R(\lambda; -A)\|_{\mathcal{L}(E_{2\alpha-1};D(A^\alpha))} |d\lambda|.$$

Now using the resolvent estimate from the previous development we get

$$\|AS(t)\|_{\mathcal{L}(E_{2\alpha-1};D(A^\alpha))} \le \frac{C}{2\pi} \int_\Gamma |\lambda|^{\frac{1}{4}+\frac{\alpha}{2}} e^{t\Re\lambda} |d\lambda| \qquad \alpha = \frac{1}{2} \ \text{or } 1.$$

Evaluating the integral we get (i). Now due to the differentiability of $S(t)$

$$S(t_2) - S(t_1) = \int_{t_1}^{t_2} \frac{dS(t)}{dt} dt = -\int_{t_1}^{t_2} AS(t) dt.$$

Hence

$$\|S(t_2) - S(t_1)\|_{\mathcal{L}(E_{2\alpha-1};D(A^\alpha))} \le \int_{t_1}^{t_2} \|AS(t)\|_{\mathcal{L}(E_{2\alpha-1};D(A^\alpha))} dt \qquad \alpha = \frac{1}{2} \ \text{or } 1.$$

Now using (i) we get (ii).

♣

Such estimates in other operator topologies can be obtained in a similar way.

4.2 Semigroup Generated By The Operator $-L$

In this section we will study the linear evolution problem obtained by perturbing a basic steady field. In other words we would like to characterize the semigroup generated by the operator $-L$ (defined in section 3.2). The first step is to obtain estimates for the resolvent $R(\lambda; -L)$.

Lemma 4.3 *The resolvent $R(\lambda; -L)$ satisfies*

$$(i) \quad \|R(\lambda; -L)\|_{\mathcal{L}(D(A^\alpha); D(A^\alpha))} \leq \frac{M}{|\lambda|}, \quad \forall \lambda \in \Sigma, \alpha \in [-1/2, 1].$$

$$(ii) \quad \|R(\lambda; -L)\|_{\mathcal{L}(E_m; D(A^\alpha))} \leq \frac{C_0}{|\lambda|^{1-(\alpha-\frac{m}{4})}} \quad \forall \lambda \in \Sigma \ and \ \alpha \in [\frac{m}{2}, 1], m = 0 \ or \ 1$$

Here

$$\Sigma = \{\lambda \in \mathbf{C}; |\lambda| > \rho; 0 < |\arg \lambda| < \frac{\pi}{2} + \delta\}$$

with ρ, δ suitably chosen positive numbers.

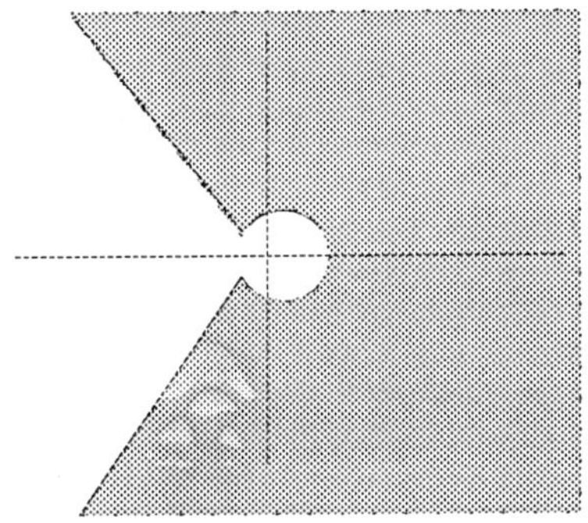

Figure 4.2.1: The region Σ

Proof

We first note that if the basic field has sufficient regularity then due to lemma 3.3 $L_u \in \mathcal{L}(B_1; B_2)$ with either,

$$(i) \quad B_1 = D(A^\alpha), B_2 = D(A^{\alpha-\frac{1}{2}}), \alpha \in [0, 1/2]$$

or

$$(ii) \quad B_1 = D(A), B_2 = E_1.$$

Hence using the estimates (of theorem 3.9) for the resolvent $\|R(\lambda; -A)\|_{\mathcal{L}(B_2; B_1)}$ we conclude that $\|L_u R(\lambda; -A)\|_{\mathcal{L}(B_2; B_2)} < 1$ for sufficiently large $|\lambda|$. This implies, $[I + L_u R(\lambda; -A)]^{-1} \in \mathcal{L}(B_2; B_2)$. Then the form $R(\lambda; -L) = R(\lambda; -A)[I + L_u R(\lambda; -A)]^{-1}$ and the estimates for the resolvent $R(\lambda; -A)$ obtained earlier in theorem 3.9 and lemma 4.3 give us the desired results.

Theorem 4.2 *The operator* $-L$ *generates a strongly continuous semigroup* $Z(t)$ *in* $D(A^\alpha)$ *which has a holomorphic extension in a sector containing the positive real axis. Moreover we have the estimates,*

$$(i) \quad \|Z(t)\|_{\mathcal{L}(D(A^\alpha);D(A^\alpha))} \leq C_0 e^{\rho t}, \quad \alpha \in [-1/2, 1]$$

$$(ii) \quad \|Z(t)\|_{\mathcal{L}(D(A^{\alpha_1});D(A^{\alpha_2}))} \leq C_1 \frac{1}{t^\alpha} e^{\rho t}, \alpha = \alpha_2 - \alpha_1 \geq 0, \alpha_1 \in [-1/2, 1]$$

$$(iii) \quad \|Z(t)\|_{\mathcal{L}(E_m;D(A^\alpha))} \leq C_2 \frac{1}{t^{\alpha - \frac{m}{4}}} e^{\rho t} \ for \ m = 0 \ or \ 1 \ and \ \alpha \in [m/2, 1].$$

Here C_0, C_1 *and* $C_2 > 0.$

Proof

From the resolvent estimate (i) of lemma 4.3 we conclude by Hille Yoshida theorem that the operator $-L$ generates a C^0 semigroup $Z(t)$ in $D(A^\alpha)$ which can be extended as a holomorphic semigroup in a sector containing the positive real axis [67]. The estimate (i) easily follows. The semigroup $Z(t)$ can be expressed by the integral

$$Z(t) = \frac{1}{2\pi i} \int_\Gamma e^{\lambda t} R(\lambda; -L) d\lambda \tag{4.1}$$

where $\Gamma = \Gamma_1 \cup \Gamma_2 \cup \Gamma_3$ with

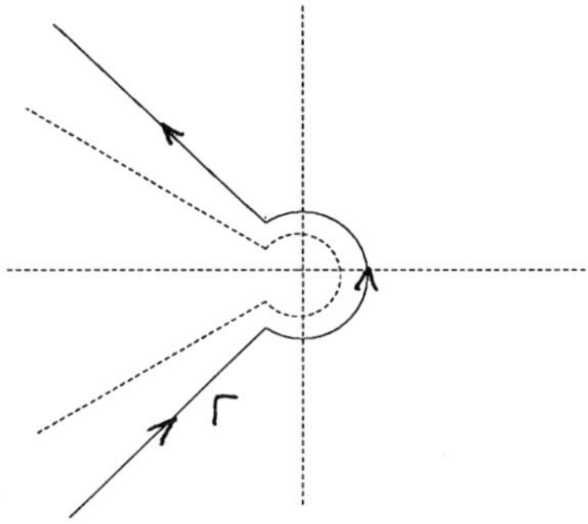

Figure 4.2.2: The Contour Γ

$$\Gamma_1 = \{xe^{-i\theta}; \rho_0 \leq x < \infty\},$$

$$\Gamma_2 = \{\rho_0 e^{i\phi}; -\theta \le \phi \le \theta\} \text{ and}$$

$$\Gamma_1 = \{x e^{i\theta}; \rho_0 \le x < \infty\}.$$

Here $\rho < \rho_0$ and $\frac{\pi}{2} < \theta < \frac{\pi}{2} + \delta$. Hence we get for $m = 0, 1$ and $\alpha \in [m/2, 1]$,

$$\|Z(t)\|_{\mathcal{L}(E_m, D(A^\alpha))} \le \frac{1}{2\pi} \int_\Gamma e^{t\Re\lambda} \frac{C_0}{|\lambda|^{1-(\alpha-m/4)}} |d\lambda|$$

This gives the estimate (iii). Proof of (ii) is similar.

$$\clubsuit$$

We note here that the analyticity of the semigroup $Z(t)$ implies the continuity in the uniform operator topology of $\mathcal{L}(E_m; D(A^\alpha)) \cap \mathcal{L}(D(A^{\alpha_1}); D(A^{\alpha_2}))$ for $t > 0$.

Theorem 4.3 *For* $t > 0$ *the operators* $S(t), Z(t) \in \mathcal{L}(E_m; D(A^\alpha)) \cap \mathcal{L}(D(A^{\alpha_1}); D(A^{\alpha_2}))$ *are compact. Here* $\alpha_1, \alpha_2 \in [-1/2, 1], \alpha_2 \ge \alpha_1, m = 0$ *or 1 and* $\alpha \in [m/2, 1]$.

Proof

As noted earlier the operator $Z(t) \in \mathcal{L}(D(A^\alpha), D(A^\alpha)), \alpha \in [-\frac{1}{2}, 1]$ is holomorphic for $t > 0$ and thus continuous in the uniform operator topology. It was established last chapter that the resolvent $R(\lambda; -L) \in \mathcal{L}(D(A^\alpha), D(A^\alpha))$ is compact for $\lambda \in \Sigma$. Note that if $\Re\lambda > \rho$ we have for $t > 0$

$$R(\lambda; -L) = \int_0^\infty e^{-\lambda t} Z(t) dt.$$

To see this multiply equation (4.1) by $e^{-\lambda t}$ and integrate with respect to t from 0 to T and use Fubini theorem to get,

$$\int_0^T e^{-\lambda t} Z(t) dt = \frac{1}{2\pi i} \int_\Gamma \frac{1}{(\mu - \lambda)} [e^{(\mu-\lambda)T} - 1] R(\mu; -L) d\mu$$

$$-\frac{1}{2\pi i} \int_\Gamma \frac{1}{(\mu - \lambda)} R(\mu; -L) d\mu + \frac{1}{2\pi i} \int_\Gamma \frac{e^{(\mu-\lambda)T}}{(\mu - \lambda)} R(\mu; -L) d\mu.$$

The first integral on the right can be seen to be equal to $R(\lambda; -L)$ by residue theorem. Let us estimate the second integral I_2

$$\|I_2\|_{\mathcal{L}(D(A^\alpha), D(A^\alpha))} \le \frac{1}{2\pi} \int_\Gamma e^{\Re(\mu-\lambda)T} \frac{\text{const.}}{|\mu - \lambda||\mu|} |d\mu| \quad \to 0 \text{ as } T \to \infty.$$

Now consider for $\lambda \in \Sigma$

$$\lambda R(\lambda; -L)Z(t) - Z(t) = \lambda \int_0^\infty e^{-\lambda s}[Z(t+s) - Z(t)]ds.$$

Let λ be real and greater than σ. Then

$$\|\lambda R(\lambda; -L)Z(t) - Z(t)\|_{\mathcal{L}(D(A^\alpha), D(A^\alpha))} \leq \int_0^\delta \lambda e^{-\lambda s}\|Z(t+s) - Z(t)\|_{\mathcal{L}(D(A^\alpha), D(A^\alpha))}ds$$

$$+ \int_\delta^\infty \lambda e^{-\lambda s}\|Z(t+s) - Z(t)\|_{\mathcal{L}(D(A^\alpha), D(A^\alpha))}ds$$

$$\leq \sup_{s \in [0,\delta]} \|Z(t+s) - Z(t)\|_{\mathcal{L}(D(A^\alpha), D(A^\alpha))} + \text{const.} \ \frac{2\lambda}{(\lambda - \sigma)} e^{\sigma(t+\delta) - \lambda \delta}.$$

Since $\delta > 0$ is arbitrary and for $t > 0$, $Z(t)$ is continuous in the uniform operator topology, we get

$$lim_{\lambda \to \infty}\|\lambda R(\lambda; -L)Z(t) - Z(t)\|_{\mathcal{L}(D(A^\alpha), D(A^\alpha))} = 0.$$

But $\forall \lambda \in \Sigma$, $R(\lambda; -L)$ is compact in $D(A^\alpha)$ and since $Z(t) \in \mathcal{L}(D(A^\alpha), D(A^\alpha))$ we have $\lambda R(\lambda; -L)Z(t)$ is compact in $D(A^\alpha)$ $\forall \lambda \in \Sigma$. Thus $Z(t)$ is compact in $D(A^\alpha)$ for $t > 0$ since it is the uniform limit of a sequence of compact operators in $D(A^\alpha)$. Note in particular that setting the basic field $\boldsymbol{U} \equiv 0$ gives us the result that $S(t), t > 0$ compact in $D(A^\alpha)$. The semigroup property $Z(t) = Z(t/2) \circ Z(t/2)$ and the fact that $Z(t/2) \in \mathcal{L}(E_m; D(A^\alpha))$ for $m = 0$ or 1 and $\alpha \in [m/2, 1]$ gives us the result that $Z(t), t > 0$ compact from E_m in to $D(A^\alpha)$. Again setting the basic velocity $\boldsymbol{U} \equiv 0$ we get $S(t), t > 0$ compact from E_m in to $D(A^\alpha)$. Compactness of the other cases can be similarly obtained.

$$\clubsuit$$

4.3   Existence Theorem For Periodic Solutions

This section will be devoted to the study of the existence of time periodic solutions to the Navier-Stokes equations. We will show in particular that each T-periodic (in time) boundary data will correspond to atleast one T-periodic (in time) generalized solution. Existence of such solutions was established by Prodi[68] in the case of two space dimensions. Yudovich[94] suggested a method that seems to be more powerful since part of the proof appears to apply to three dimensional flows. We will develop and elaborate his method in this section. The steps involved are the following. We first show that a T-periodic field is a weak solution in $\boldsymbol{R}$ if and only if it is a weak solution in $[0, T]$. We then show that there is a weak solution in $[0, T]$ which returns to its original value after time T. If this solution is single valued in the interval $[0, T]$ then we can obtain the (periodic) weak solution in $\boldsymbol{R}$ by periodic extension. We note here that the Hopf class weak solution [38] has been proven to be single valued only for two dimensional flows.

Let us consider the problem 2.1 (without the initial data) with the boundary data satisfying

$$\boldsymbol{u}_b \in C(\boldsymbol{R}; \overset{o}{\boldsymbol{H}}{}^{1/2}(\partial\Omega)) \text{ and } \frac{\partial \boldsymbol{u}_b}{\partial t} \in L^2_{\text{loc}}(\boldsymbol{R}; \overset{o}{\boldsymbol{H}}{}^{1/2}(\partial\Omega)). \tag{4.2}$$

Moreover $\boldsymbol{u}_b$ and $\frac{\partial \boldsymbol{u}_b}{\partial t}$ are T-periodic in time. Here $C(\boldsymbol{R}; X)$ denotes continuous and bounded functions from $\boldsymbol{R}$ in to X.

The following theorem provides the suitable Hopf type extension of the boundary data (compare with the corresponding result for the stationary basic flow in chapter 3).

Theorem 4.4 *For all $\delta > 0$, there exists a continuous linear map*

$$\Lambda_\delta : C(\boldsymbol{R}; \overset{o}{\boldsymbol{H}}{}^{1/2}(\partial\Omega)) \to C(\boldsymbol{R}; \boldsymbol{H}^1(\Omega))$$

such that $\forall \boldsymbol{u}_b \in C(\boldsymbol{R}; \overset{o}{\boldsymbol{H}}{}^{1/2}(\partial\Omega))$,

$$(i) \quad div \cdot \Lambda_\delta \boldsymbol{u}_b = 0 \text{ in } \Omega \times \boldsymbol{R}$$

$$(ii) \quad \gamma_0 \Lambda_\delta \boldsymbol{u}_b = \boldsymbol{u}_b \text{ on } \partial\Omega \times \boldsymbol{R}$$

where γ_0 is the first-trace operator.

$$(iii) \quad |b(\boldsymbol{v}, \Lambda_\delta \boldsymbol{u}_b, \boldsymbol{v})| \leq \delta \|\boldsymbol{v}\|_V^2 \|\boldsymbol{u}_b\|_{C(\boldsymbol{R}; \overset{o}{\boldsymbol{H}}{}^{1/2}(\partial\Omega))}, \qquad \forall \boldsymbol{v} \in \boldsymbol{V}$$

and if $\frac{\partial}{\partial t} \boldsymbol{u}_b \in L^2_{loc}(\boldsymbol{R}; \overset{o}{\boldsymbol{H}}{}^{1/2}(\partial\Omega))$ then

$$(iv) \quad \frac{\partial}{\partial t} \Lambda_\delta \boldsymbol{u}_b \in L^2_{loc}(\boldsymbol{R}; \boldsymbol{H}^1(\Omega)).$$

Proof

Let us consider the problem

$$-\Delta\phi(x,t) + \nabla q(x,t) = 0, \quad (x,t) \in \Omega \times \boldsymbol{R}$$

$$\nabla \cdot \phi(x,t) = 0, \quad (x,t) \in \Omega \times \boldsymbol{R}$$

$$\phi(x,t) = u_b(x,t), \quad (x,t) \in \partial\Omega \times \boldsymbol{R}.$$

A variational formulation of the above problem gives the existence of a unique solution $\phi(x,t) \in C(\boldsymbol{R}; \boldsymbol{H}^1(\Omega))$ such that

$$\gamma_0\phi = u_b \in C(\boldsymbol{R}; \overset{\circ}{\boldsymbol{H}}{}^{1/2}(\partial\Omega)).$$

Since for each t, the correspondence $u_b \to \phi$ is linear we get $\phi_t \in L^2_{\text{loc}}(\boldsymbol{R}; \boldsymbol{H}^1(\Omega))$. Note also that if u_b is T-periodic in time so is ϕ.

From the isomorphism property of the curl operator established in chapter 3 it is clear that,

Lemma 4.4

$$curl\ C(\boldsymbol{R}; \boldsymbol{H}^2(\Omega) \cap \boldsymbol{H}_0) = curl\ C(\boldsymbol{R}; \boldsymbol{H}^2(\Omega))$$

and curl is an isomorphism from $C(\boldsymbol{R}; \boldsymbol{H}^2(\Omega) \cap \boldsymbol{H}_0)$ *on to* $curlC(\boldsymbol{R}; \boldsymbol{H}^2(\Omega))$.

We then deduce from lemmas 4.4 and 3.1,that there exists $\psi \in C(\boldsymbol{R}; \boldsymbol{H}^2(\Omega) \cap \boldsymbol{H}_0)$ such that $\phi(x,t) = curl\ \psi(x,t)$.

The following two results are proved in Hopf[38] :

Proposition 4.1 *Let* $\rho(x) = dist(x, \partial\Omega)$. *Then for all* $\epsilon > 0$, *there exists* $\theta_\epsilon \in C^2(\bar{\Omega})$ *such that,*

(i) $\theta_\epsilon = 1$ *in some neighborhood of* $\partial\Omega$

(ii) $\theta_\epsilon = 0$ *if* $\rho(x) > 2\delta(\epsilon) = 2\exp(-1/\epsilon)$ *and*

(iii) $|\frac{\partial}{\partial x^k}\theta_\epsilon| \leq \epsilon/\rho(x)$ *if* $\rho(x) \leq 2\delta(\epsilon), k = 1, \cdots, n.$

Proposition 4.2 *There exists a positive constant C_1 depending on Ω such that*

$$\|\frac{1}{\rho}v\|_{L^2(\Omega)} \le C_1\|v\|_{H^1(\Omega)} \text{ for all } v \in \boldsymbol{H}_0^1(\Omega).$$

Now set $\Lambda_\delta u_b = \text{curl } (\theta_\epsilon(x)\psi(x,t))$. We then immediately get properties (i),(ii) and (iv). To get property (iii), we first note that due to Sobolev embedding theorem $\psi(x,t)$ is continuous in (x,t). Hence for each t,

$$|b(v,\Lambda_\delta u_b,v)| = |-b(v,v,\Lambda_\delta u_b)| = |\sum_{i,j=1}^{n} \int_\Omega v_i(\Lambda_\delta u_b)_j \frac{\partial v_j}{\partial x^i}dx|$$

$$\le [\sum_{i,j=1}^{n} \|v_i(\Lambda_\delta u_b)_j\|_{L^2(\Omega)}^2]^{1/2}\|v\|_V \quad \forall v \in \boldsymbol{V}.$$

Estimating the first term with the help of propositions 4.1 and 4.2 we get (iii).

♣

Let us introduce a change of variable of the form $u = v + \Lambda_\delta u_b$ in the equation (2.1) to get,

$$v_t + (v \cdot \nabla)v + (v \cdot \nabla)(\Lambda_\delta u_b) + (\Lambda_\delta u_b) \cdot \nabla v = -\nabla p + \nu\Delta v + f \text{ in } \Omega \times \boldsymbol{R}$$

$$\nabla \cdot v = 0 \text{ in } \Omega \times \boldsymbol{R} \tag{4.3}$$

$$v(x,t) = 0 \quad (x,t) \in \partial\Omega \times \boldsymbol{R} \text{ and}$$

$$v(x,t+T) = v(x,t) \quad (x,t) \in \Omega \times \boldsymbol{R}$$

Here

$$f = -\frac{\partial(\Lambda_\delta u_b)}{\partial t} - (\Lambda_\delta u_b) \cdot \nabla(\Lambda_\delta u_b) + \nu\Delta(\Lambda_\delta u_b)$$

and is T-periodic in time. Note that

$$\Delta(\Lambda_\delta u_b) \in C(\boldsymbol{R}; \boldsymbol{V}')$$

and

$$(\Lambda_\delta u_b) \cdot \nabla(\Lambda_\delta u_b) \in C(\boldsymbol{R}; L^{3/2}(\Omega)).$$

But $L^{3/2}(\Omega) = (L^3(\Omega))'$ and by the Sobolev embedding theorem

$$(L^3(\Omega))' \subset (\boldsymbol{H}_0^1(\Omega))' = \boldsymbol{H}^{-1}(\Omega) \subset \boldsymbol{V}'$$

Hence $(\Lambda_\delta u_b) \cdot \nabla(\Lambda_\delta u_b) \in C(\mathbf{R}; \mathbf{V}')$. Noting that

$$\frac{\partial}{\partial t}(\Lambda_\delta u_b) \in L^2_{\mathrm{loc}}(\mathbf{R}; \mathbf{H}^1(\Omega))$$

we write $f \in L^2_{\mathrm{loc}}(\mathbf{R}; \mathbf{V}')$.

Let us now consider the problem of finding a solution $v \in L^2_{\mathrm{loc}}(\mathbf{R}; \mathbf{V})$ such that $\forall w \in \mathbf{V}$,

$$\frac{d}{dt}(v, w) + \nu a(v, w) + b(v, v, w) + b(v, \Lambda_\delta u_b, w)$$

$$b(\Lambda_\delta u_b, v, w) = < f, w >_{V' \times V}, \text{ in } \mathcal{D}(\mathbf{R})' \tag{4.4}$$

$$\text{with } v(t) = v(t + T) \quad \in \mathbf{V}' \quad \forall t \in \mathbf{R}.$$

Here $\mathcal{D}(\mathbf{R})'$ is the space of distributions on $\mathbf{R}$. We will first interpret the sense in which such a solution satisfies the periodic condition. Note that the trilinear form $b(\cdot, \cdot, \cdot)$ is continuous on $\mathbf{V} \times \mathbf{V} \times \mathbf{V}$ and thus it can be associated (using the Riesz representation theorem) with a duality pairing on $\mathbf{V}' \times \mathbf{V}$. Hence for almost all $t \in \mathbf{R}$ we have $\forall w \in \mathbf{V}$,

$$< B(v, v), w >_{V' \times V} = b(v, v, w),$$

$$< B(\Lambda_\delta u_b, v), w >_{V' \times V} = b(\Lambda_\delta u_b, v, w), \text{ and}$$

$$< B(v, \Lambda_\delta u_b), w >_{V' \times V} = b(v, \Lambda_\delta u_b, w).$$

Since $v \in L^2_{\mathrm{loc}}(\mathbf{R}; \mathbf{V})$ and $\Lambda_\delta u_b \in C(\mathbf{R}; \mathbf{H}^1(\Omega))$ we get $B(v, v) \in L^1_{\mathrm{loc}}(\mathbf{R}; \mathbf{V}')$ and

$$B(v, \Lambda_\delta u_b), B(\Lambda_\delta u_b, v) \in L^2_{\mathrm{loc}}(\mathbf{R}; \mathbf{V}').$$

Note that $Av \in L^2_{\mathrm{loc}}(\mathbf{R}; \mathbf{V}')$ due to the isomorphism $A : \mathbf{V} \to \mathbf{V}'$. Now write $\forall w \in \mathbf{V}$,

$$\frac{d}{dt} < v, w >_{V' \times V} = < f - \nu Av - B_\delta(v), w >_{V' \times V} \quad \in \mathcal{D}(\mathbf{R})' \tag{4.5}$$

where $B_\delta(v)$ is the sum of the three bilinear terms $B(\cdot, \cdot)$. This implies [54] that

$$v' = f - \nu Av - B_\delta(v) \in L^1_{\mathrm{loc}}(\mathbf{R}; \mathbf{V}')$$

and hence $v \in C(\mathbf{R}; \mathbf{V}')$. Let us now establish the existence theorem.

64

Theorem 4.5 *Let $\Omega \subset R^2$ be a bounded open set of class C^2 and u_b satisfies the conditions in equation (4.2). Then there exists atleast one T-periodic solution v to the problem 2.1 such that, $v \in L^2_{loc}(R; V) \cap C(R; H)$ and,*

$$\int_{-\infty}^{+\infty} [-(v(\tau), w'(\tau))_{L^2(\Omega)} + \, < B_\delta(v), w >_{V' \times V}$$

$$+ \, a(v, w) - \, < f(\tau), w(\tau) >_{V' \times V}] d\tau = 0 \quad \forall w \in \mathcal{Y} \tag{4.6}$$

where

$$\mathcal{Y} = \{ w \in C(R; V); w' \in L^2(R; H); \text{support of } w \subset\subset R \}.$$

Proof

Let us first show that a necessary and sufficient condition for v to be a T-periodic weak solution in R is that it be a weak solution in the interval $[0, T]$. Note that if v is a Hopf-class weak solution in R then taking $\psi_0 \in \mathcal{Y}$ with support contained in $[0, T]$ we get the weak solution in this interval satisfying,

$$\int_0^T [[\cdot, \cdot]](t) dt = 0, \forall \psi_0 \in \mathcal{Y}, \; \text{supp} \psi_0 \in [0, T] \tag{4.7}$$

Here $[[\cdot, \cdot]]$ represents the terms inside the square bracket of the equation (4.6). On the other hand if $v(x, t)$ is a weak solution in $[0, T]$ then

$$< v(T), \psi(T) >_{V' \times V} - \, < v(0), \psi(0) >_{V' \times V} +$$

$$\int_0^T [[\cdot, \cdot]](t) dt = 0, \forall \psi \in \mathcal{Y}.$$

Using the periodicity of v and the data we can rewrite this as,

$$< v((k+1)T), \psi_k((k+1)T) >_{V' \times V} - \, < v(kT), \psi_k(kT) >_{V' \times V}$$

$$+ \int_{kT}^{(k+1)T} [[\cdot, \cdot]](t) dt = 0, \forall \psi_k \in \mathcal{Y}, k \in Z$$

with $\psi_k(t) = \psi(t - kT)$. Summing over k we get (4.6).

We will now show that within the frame work of the Hopf class weak solutions in the interval $[0, T]$ there is a solution for which the value at time T coincide with the initial data.

Let us consider a sequence of vector fields $\{\psi_k(x)\}_{k=1}^{\infty}, \psi_k \in j(\Omega), \forall k$ that is free and total in $\boldsymbol{V}$. Taking the Galerkin projection of the weak form (4.5) we get,

$$(v'_m, \psi_k) + \nu a(v_m, \psi_k) + \langle B_\delta(v_m), \psi_k \rangle = \langle f, \psi_k \rangle, k = 1, \cdots, m, \tag{4.8}$$

where $v_m(x) = \sum_{k=1}^{m} g_k(t)\psi_k(x)$.

Let us consider the initial value problem associated with this system of ordinary differential equations for the evolution of $\{g_k(t)\}_{k=1}^{m}$. Let the initial data be $u_{0m} = \sum_{k=1}^{m} a_k \psi_k(x)$. Then there exists a unique solution $\{g_1(t), \cdots, g_m(t)\}$ that continuously depends on the initial data $\{a_1, \cdots, a_m\}$ [14]. Let the correspondence between the solution and the data be denoted by a map $\mathcal{F}$ in $\boldsymbol{R}^m$ such that

$$g_i(t) = \mathcal{F}_i(a_1, \cdots, a_m; t), i = 1, \cdots, m.$$

We will now show that $\mathcal{F}(\cdots; T)$ (which is continuous) maps a ball in $\boldsymbol{R}^m$ in to itself. Multiply equation (4.8) by g_k and sum over k to obtain,

$$\frac{d}{dt}\|v_m\|_{L^2(\Omega)}^2 + 2\nu\|v_m\|_V^2 + 2b(v_m, \Lambda_\delta u_b, v_m)$$

$$= 2\langle f, v_m \rangle \quad \leq 2\|f\|_{V'}\|v_m\|_V.$$

Now using the properties of the Hopf -extension operator and the Young's inequality, we get,

$$\frac{d}{dt}\|v_m\|_{L^2(\Omega)}^2 + (\nu - \delta)\|v_m\|_V^2 \leq \frac{1}{(\nu - \delta)}\|f\|_{V'}^2. \tag{4.9}$$

Now using Poincare's lemma we get,

$$\frac{d}{dt}\|v_m\|_{L^2(\Omega)}^2 + (\nu - \delta)\mu_1\|v_m\|_{L^2(\Omega)}^2 \leq \frac{1}{(\nu - \delta)}\|f\|_{V'}^2.$$

Using Gronwall's lemma [91] we obtain

$$\|v_m(t)\|_{L^2(\Omega)}^2 \leq \|v_m(0)\|_{L^2(\Omega)}^2 - \{\|v_m(0)\|_{L^2(\Omega)}^2(1 - e^{-\beta t})$$

$$- \frac{1}{\nu - \delta}\int_0^t e^{-\beta(t-\tau)}\|f(\tau)\|_{V'}^2 d\tau\} \tag{4.10}$$

Here $\beta = (\nu - \delta)\mu_1$.

Let us consider an initial data contained in the closed ball $B_R(0)$ of radius R in $\boldsymbol{H}$ centered at the origin:

$$\|\boldsymbol{v}_m(0)\|_{L^2(\Omega)} \leq R.$$

Then from the estimate (4.10) we can conclude that for sufficiently large R

$$\|\boldsymbol{v}_m(T)\|_{L^2(\Omega)} \leq R.$$

This implies that $\mathcal{F}(\cdot;T)$ continuously maps a closed ball in $\boldsymbol{R}^m$ in to itself and hence by Brower's fixed point theorem there exists a fixed point $\{a_k^p\}_{k=1}^m$ such that

$$a_k^p = \mathcal{F}_k(a_1^p, \cdots, a_m^p; T), \quad k = 1, \cdots, m.$$

Let us now establish the limit of these solution.

The estimate (4.10) implies that $\|\boldsymbol{v}_m(x,t)\|_{L^2(\Omega)}$ is uniformly bounded (independent of t and m). Integrating the equation (4.9) and using the above uniform bound we can deduce that $\|\boldsymbol{v}_m(x,t)\|_{L^2(0,T;V)}$ is uniformly bounded (independent of m). Thus we can extract a subsequence from $\{\boldsymbol{v}_m\}$ that converges weakly in $L^2(0,T;V)$ and weak-star in $L^\infty(0,T;\boldsymbol{H})$. Moreover as in [38] we can show that $\boldsymbol{v}_m$ belongs to a compact set in $L^2(0,T;\boldsymbol{H})$. This allows us to extract a subsequence which converges strongly in $L^2(0,T;\boldsymbol{H})$.

The following lemma will be used to establish the limit to the Navier Stokes equations.

Lemma 4.5 *Let $\{\boldsymbol{v}_k\}$ be a sequence of vectorfields that converges weakly in $L^2(0,T;\boldsymbol{V})$, weak star in $L^\infty(0,T;\boldsymbol{H})$ and strongly in $L^2(0,T;\boldsymbol{H})$ then the following limits are obtained $\forall \boldsymbol{w} \in \mathcal{Y}.$*

$$(i) \lim_{k\to\infty} \int_0^T b(\boldsymbol{v}_k, \boldsymbol{v}_k, \boldsymbol{w})dt = \int_0^T b(\boldsymbol{v}, \boldsymbol{v}, \boldsymbol{w})dt.$$

$$(ii) \lim_{k\to\infty} \int_0^T b(\boldsymbol{v}_k, \Lambda_\delta \boldsymbol{u}_b, \boldsymbol{w})dt = \int_0^T b(\boldsymbol{v}, \Lambda_\delta \boldsymbol{u}_b, \boldsymbol{w})dt.$$

$$(iii) \lim_{k\to\infty} \int_0^T b(\Lambda_\delta \boldsymbol{u}_b, \boldsymbol{v}_k, \boldsymbol{w})dt = \int_0^T b(\Lambda_\delta \boldsymbol{u}_b, \boldsymbol{v}, \boldsymbol{w})dt.$$

$$(iv) \lim_{k\to\infty} \int_0^T (\boldsymbol{v}_k, \boldsymbol{w}')_{L^2(\Omega)}dt = \int_0^T (\boldsymbol{v}, \boldsymbol{w}')_{L^2(\Omega)}dt.$$

Proof

Recall the estimate for the trilinear form [45]

$$|b(\boldsymbol{v}, \boldsymbol{u}, \boldsymbol{w})| \leq C\|\boldsymbol{v}\|_{L^2(\Omega)}^{1/4}\|\boldsymbol{v}\|_{H^1(\Omega)}^{3/4}\|\boldsymbol{u}\|_{H^1(\Omega)}\|\boldsymbol{w}\|_{L^2(\Omega)}^{1/4}\|\boldsymbol{w}\|_{H^1(\Omega)}^{3/4}$$

$$\forall \boldsymbol{v}, \boldsymbol{u}, \boldsymbol{w} \in \boldsymbol{H}^1(\Omega).$$

Hence writing

$$b(\boldsymbol{v}, \boldsymbol{v}, \boldsymbol{w}) - b(\boldsymbol{v}_k, \boldsymbol{v}_k, \boldsymbol{w}) = b(\boldsymbol{v} - \boldsymbol{v}_k, \boldsymbol{v}, \boldsymbol{w}) + b(\boldsymbol{v}_k, \boldsymbol{v} - \boldsymbol{v}_k, \boldsymbol{w})$$

we estimate

$$|\int_0^T [b(\boldsymbol{v}, \boldsymbol{v}, \boldsymbol{w}) - b(\boldsymbol{v}_k, \boldsymbol{v}_k, \boldsymbol{w})dt|$$

$$\leq C_1\|\boldsymbol{v} - \boldsymbol{v}_k\|_{L^2(0,T;H)}^{1/4}\|\boldsymbol{v} - \boldsymbol{v}_k\|_{L^2(0,T;V)}^{3/4}\|\boldsymbol{v}\|_{L^2(0,T;V)}\|\boldsymbol{w}\|_{C(0,T;H^1(\Omega)}$$

$$+C_2\|\boldsymbol{v} - \boldsymbol{v}_k\|_{L^2(0,T;H)}^{1/4}\|\boldsymbol{v} - \boldsymbol{v}_k\|_{L^2(0,T;V)}^{3/4}\|\boldsymbol{v}_k\|_{L^2(0,T;V)}\|\boldsymbol{w}\|_{C(0,T;H^1(\Omega)}.$$

Since $\|\boldsymbol{v} - \boldsymbol{v}_k\|_{L^2(0,T;V)}$, $\|\boldsymbol{v}_k\|_{L^2(0,T;V)}$ and $\|\boldsymbol{w}\|_{C(0,T;H^1(\Omega)}$ are bounded and

$$\lim_{k\to\infty} \|\boldsymbol{v} - \boldsymbol{v}_k\|_{L^2(0,T;H)} = 0$$

we get (i). To obtain (ii) note that

$$\int_0^T [b(\boldsymbol{v}_k, \Lambda_\delta u_b, \boldsymbol{w}) - b(\boldsymbol{v}, \Lambda_\delta u_b, \boldsymbol{w})]dt$$

$$\leq C_3\|\Lambda_\delta u_b\|_{C(0,T;H^1(\Omega)}\|\boldsymbol{w}\|_{C(0,T;V)}\|\boldsymbol{v} - \boldsymbol{v}_k\|_{L^2(0,T;H)}^{1/4}\|\boldsymbol{v} - \boldsymbol{v}_k\|_{L^{6/7}(0,T;V)}^{4/3}$$

and tends to zero due to the strong convergence of $\boldsymbol{v}_k$ in $L^2(0, T; \boldsymbol{H})$ and since $\boldsymbol{v}_k$ and $\boldsymbol{v}$ are bounded in $L^2(0, T; \boldsymbol{H})$. Proof of (iii) is similar to that for (ii). Note that

$$\lim_{k\to\infty} \int_0^T (\boldsymbol{v} - \boldsymbol{v}_k, \boldsymbol{w}')_{L^2(\Omega)}dt = 0$$

since $\boldsymbol{w}' \in L^2(0, T; \boldsymbol{H})$ and $\boldsymbol{v}_k$ converges strongly in $L^2(0, T; \boldsymbol{H})$.

♣

68

Let us now describe the convergence of the end points. Note that,

$$< v'_m, \psi_k >_{V' \times V} = < f - B_\delta(v_m) - \nu A v_m, \psi_k >_{V' \times V}, \forall k.$$

This implies

$$v'_m = f - B_\delta(v_m) - \nu A v_m \in L^1(0, T; V')$$

and hence $v_m(x,t) \in C([0,T]; V')$ (See remark below.). Thus, the end points satisfy

$$v_m(x, 0) = v_m(x, T) \in V'.$$

Now note that $\forall t \in [0, T]$, a subsequence $v_{m'}(x,t) \to v(x,t)$ weakly in V'. Thus,

$$v(x, 0) = v(x, T) \quad \in V'.$$

Using lemma 4.5 and the convergence of the end points we can conclude that the limit solution satisfies 4.6.

Remark: In two dimensions we actually have $v_m \in C([0,T]; H)$.

♣

4.3.1 Karman Vortex Shedding

Karman vortex shedding would probably the most spectacular (mathematical) demonstration of the *Hopf bifurcation* phenomena, to a practical fluid dynamical problem. In this context we will briefly note a result of Babenko [6] concerning the time priodic viscous flow past a three dimensional obstacle. The existence theorem proven in the previous section for bounded domains, holds for arbitrary Reynolds numbers. In the Babenko's study however, the periodic solution is found via Hopf bifurcation from the Leray-Finn-Babenko steady solution as the Reynold's number increases past the critical value. Here the critical Reynold's number is defined (in Theorem 3.12) as the largest value for which the discrete part of the Babenko spectrum is contained in the right half plane. Since the dependence of these eigenvalues on the Reynold's number is analytic (theorem 3.12), as the Reynold's number increases past the critical value, a number of eigenvalues will cross the imaginary axis. Hence, at $R_e = R_{ec}$ we can certainly find a number of them on the imaginary axis and the rest on the right half plane. Babenko considers the problem of finding $(u, p) : \Omega \times R \to R^3 \times R$ such that

$$u(x, t) = u_L(x, R_e) + R_e^{-1} v(x, t; R_e)$$

$$p(x, t) = p_L(x, R_e) + R_e^{-2} q(x, t; R_e)$$

such that (after the substitution in to the time dependent Navier Stokes problem for the exterior domain) v and p satisfy

$$R_e \frac{\partial v}{\partial t} + R_e\{(u_L \cdot \nabla)v + (v \cdot \nabla)u_L\} + v \cdot \nabla v = \Delta v - \nabla q \text{ in } \Omega$$

$$\nabla \cdot v = 0 \quad \text{in } \Omega$$

$$v|_{\partial \Omega} = 0 \text{ and } v \to 0 \text{ as } |x| \to \infty.$$

Assumption 4.1 *The linear homogeneous problem of finding $(w, p) : \Omega \to R^3 \times R$ such that*

$$R_{ec}\{(u_L \cdot \nabla)w + (w \cdot \nabla)u_L\} = \Delta w - \nabla p \text{ in } \Omega$$

$$\nabla \cdot w = 0 \quad \text{in } \Omega$$

$$w|_{\partial \Omega} = 0 \text{ and } w \to 0 \text{ as } |x| \to \infty$$

has only trivial solution $(w, p) \equiv 0$.

This assumption is equivalent to saying that at $R_e = R_{ec}$, origin is not an eigenvalue $(0 \notin \{\lambda_i\})$ of the Babenko spectrum. Note that by taking inner product with w we can obtain a variational formulation of this linear problem,

$$R_{ec}[b(u_L, w, w) + b(w, u_L, w)] + a(w, w) = 0.$$

Now, using the decay estimates for u_L

$$|b(u_L, w, w)| \leq C \int_\Omega |\nabla w| \frac{|w|}{|x|} dx$$

$$\leq C\|\nabla w\|_{L^2(\Omega)}\|\frac{|w|}{|x|}\|_{L^2(\Omega)}$$

$$\leq C_1\|\nabla w\|_{L^2(\Omega)}^2.$$

Similarly using the decay of ∇u_L we estimate,

$$|b(w, u_L, w)| \leq C_2\|\frac{|w|}{|x|}\|_{L^2(\Omega)}^2$$

$$\leq C_3\|\nabla w\|_{L^2(\Omega)}^2.$$

Hence we have

$$\{-\boldsymbol{R}_{ec}(C_1 + C_3) + 1\}\|\nabla w\|_{L^2(\Omega)}^2 = 0.$$

Therefore $w = 0$ provided that $\boldsymbol{R}_{ec} < (C_1 + C_3)^{-1}$.

Assumption 4.2 *At* $\boldsymbol{R}_e = \boldsymbol{R}_{ec}$, *the eigenvalues on the imaginary axis are of the form*

$$\pm ic_0, \pm ic_1, \cdots, \pm ic_{k-1} \text{ with } c_0 \neq c_j, \; 1 \leq j \leq k-1$$

where c_j, $j = 0, \cdots, k-1$ *are real. Moreover*

$$nc_0 \neq c_j, j = 1, \cdots, k-1 \text{ for any integer } n > 1$$

Theorem 4.6 *:Babenko[6]*

Let the assumptions 4.1 and 4.2 are satisfied. Then, in the neighborhood of the stationary solution u_L, $\exists$ *a one parameter family of periodic solutions*

$$u(x,t) = u_L(x, \boldsymbol{R}_e) + \boldsymbol{R}_e^{-1}v(x,t; \boldsymbol{R}_e)$$

$$p(x,t) = p_L(x, \boldsymbol{R}_e) + \boldsymbol{R}_e^{-2}q(x,t; \boldsymbol{R}_e)$$

which depend analytically on time.

Remarks:

(i) For each t, the periodic solution $u \in \boldsymbol{V}_0(\Omega)$.

(ii) The actual statement of this theorem includes a Fredholm alternative type condition involving the adjoint eigenfunction of the Babenko- spectral problem and certain coefficients of the power series expansion used in the construction of this periodic solution. We refer the readers to [6] for details.

4.4 Characterization Of The Monodromy Operator

In this section we will analyze the evolution problem obtained by linearizing the Navier-Stokes equations about a smooth time dependent basic field. To obtain this linear problem we simply need to drop the nonlinear term in equation (2.2). We will study in particular the evolution operator $Z(\cdot,\cdot)$ associated with this problem. When the basic flow is time periodic or stationary, the evolution operator (which is then called the Monodromy operator) possesses certain desirable properties towards the subsequent development of the invariant manifold theorems. We will first define the evolution operator corresponding to a general basic field and study its regularity and compactness properties. In the resolution of the nonlinear hydrodynamic semigroup (considered in the next chapter) the initial data will be prescribed in $D(A^\alpha), \alpha \in [1/2, 1]$. Thus we need to characterize $Z(\cdot,\cdot)$ as an element in $\mathcal{L}(D(A^\alpha); D(A^\alpha))$. Moreover the bilinear operator characterizing the inertia terms has the property

$$B(\cdot,\cdot) : D(A^{1/2}) \times D(A^{1/2}) \to D(A^{-1/4}) \text{ and}$$

$$B(\cdot,\cdot) : D(A) \times D(A) \to E_1$$

Hence we need to extend $Z(\cdot,\cdot)$ as an element in $\mathcal{L}(D(A^{-1/4}); D(A^{1/2})) \cap \mathcal{L}(E_1; D(A))$. The compactness property of $Z(\cdot,\cdot)$ enables us to establish a spectral theorem for this operator. We will then specialize our analysis to the case where the basic flow is periodic in time and note the additional spectral properties of the evolution (monodromy) operator. Unlike the previous work on the evolution operator [40, 95], a general functional setting is provided here. The developments of this section deal with the case of bounded domains.

Let us consider the following linear problem of evolution:

Problem 4.1 *Find* $v \in C((0,\infty); D(A^\alpha)) \cap C^1((0,\infty); D(A^{\alpha-1}))$ *such that*

$$\frac{dv}{dt} + Av + L_u(t)v = 0, \quad t > 0 \tag{4.11}$$

$$v(0) = v_0 \in D(A^\beta)$$

with $\alpha \in [0,1]$ *and* $\beta \in\,]-1/2, 1]$.

Here $L_u(t)$ is characterized (using the Riesz representation theorem) by the duality pairing,

$$< L_u(t)\boldsymbol{v}, \boldsymbol{w} >_{V'\times V} = b(\boldsymbol{U}(t), \boldsymbol{v}, \boldsymbol{w}) + b(\boldsymbol{v}, \boldsymbol{U}(t), \boldsymbol{w}), \forall \boldsymbol{w} \in V$$

and $\boldsymbol{U}(t)$ is the basic field that is sufficiently regular.

Let us now consider an integral representation for $\boldsymbol{v}(t)$:

Problem 4.2 *Find* $\boldsymbol{v} \in C((0,\infty); D(A^\alpha))$ *such that,*

$$\boldsymbol{v}(t) = S(t)\boldsymbol{v_0} - \int_0^t S(t-\tau)L_u(\tau)\boldsymbol{v}(\tau)d\tau \tag{4.12}$$

$$\boldsymbol{v}(0) = \boldsymbol{v_0} \in D(A^\beta),$$

with $\alpha \in [0,1]$ *and* $\beta \in]-1/2, 1]$.

Theorem 4.7 *Let the basic field satisfies* $\boldsymbol{U} \in C([0,\infty); \boldsymbol{H}^2(\Omega))$. *Then the problem 4.2 resolves the problem 4.1.*

Proof

First note that (4.12) formally satisfies (4.11). Moreover, due to the properties of the Stokes semigroup established in chapter 4 we have

$$S(t)\boldsymbol{v_0} \in \boldsymbol{C}((0,\infty); D(A^\alpha)) \cap \boldsymbol{C}^1((0,\infty); D(A^{\alpha-1})).$$

Hence we only need to establish that

$$\int_0^t S(t-\tau)L_u(\tau)\boldsymbol{v}(\tau)d\tau \in \boldsymbol{C}((0,\infty); D(A^\alpha)) \cap \boldsymbol{C}^1((0,\infty); D(A^{\alpha-1})).$$

The following two lemmas establish this and other properties of this integral.

Lemma 4.6 *Let* $\mathcal{K}$ *be a linear operator defined as*

$$[\mathcal{K}\boldsymbol{v}](t) = - \int_0^t S(t-\eta)L_u(\eta)\boldsymbol{v}(\eta)d\eta.$$

Then for sufficiently small T_1, $\mathcal{K}$ *is a contraction in* $C([0,T_1); D(A^\alpha)), \alpha \in [0,1]$. *Moreover,* $\mathcal{K}$ *can be extended as a contraction in the Banach space* $\mathcal{B}$ *defined as:*

$$\mathcal{B} = \{t \rightarrow \boldsymbol{v}(t); \boldsymbol{v}(t) \text{ continuous in } D(A^\alpha) \text{ for } t \in]0, T_1]$$

$$\textit{and } t^{\gamma}\boldsymbol{v}(t) \textit{ bounded in } D(A^{\alpha}), \alpha \in [0,1]\}$$

$$\textit{with norm } \|\boldsymbol{v}\|_{\mathcal{B}} = \sup_{t\in(0,T_1)} \|t^{\gamma}\boldsymbol{v}(t)\|_{D(A^{\alpha})} \textit{ where}$$

γ depends on α and β.

Proof

let us consider

$$[\mathcal{K}\boldsymbol{v}](t+h) - [\mathcal{K}\boldsymbol{v}](t) = -\int_0^t [S(t+h-\tau) - S(t-\tau)]L_u(\tau)\boldsymbol{v}(\tau)d\tau$$

$$-\int_t^{t+h} S(t+h-\tau)L_u(\tau)\boldsymbol{v}(\tau)d\tau.$$

From the lemma 3.3 we can deduce that

$$L_u : C([0,\infty); D(A^{\alpha})) \rightarrow C([0,\infty); D(A^{\alpha-1/2})) \quad \alpha \in [0, 1/2],$$

and

$$L_u : C([0,\infty); D(A^{\alpha})) \rightarrow C([0,\infty); E_{2\alpha-1}) \quad \alpha = \frac{1}{2} \text{ or } 1.$$

Now using the estimates obtained in chapter 4 for the Stokes semigroup

$$S(t) \in \mathcal{L}(D(A^{\alpha}); D(A^{\alpha})) \cap \mathcal{L}(E_1; D(A)) \cap \mathcal{L}(D(A^{\beta}); D(A^{\alpha}))$$

we have for $\boldsymbol{v} \in C([0, T_1); D(A^{\alpha}))$,

$$\|[\mathcal{K}\boldsymbol{v}](t+h) - [\mathcal{K}\boldsymbol{v}](t)\|_{D(A^{\alpha})} \leq C_1 \int_0^t \{\frac{1}{(t-\tau)^{\delta}} - \frac{1}{(t+h-\tau)^{\delta}}\}\|\boldsymbol{v}(\tau)\|_{D(A^{\alpha})}d\tau$$

$$+C_2 \int_t^{t+h} \frac{1}{(t+h-\tau)^{\delta}}\|\boldsymbol{v}(\tau)\|_{D(A^{\alpha})}d\tau,$$

with $\delta = 1/2$ for $\alpha \in [0, 1/2]$ and $\delta = (1+2\alpha)/4$ for $\alpha = 1/2$ or 1. This gives us

$$\|[\mathcal{K}\boldsymbol{v}](t+h) - [\mathcal{K}\boldsymbol{v}](t)\|_{D(A^{\alpha})} \leq C_3\{h^{1-\delta} + t^{1-\delta} - (t+h)^{1-\delta}\}\|\boldsymbol{v}\|_{C([0,T_1);D(A^{\alpha}))}.$$

Right hand side $\rightarrow 0$ as $h \rightarrow 0$ and hence we conclude that

$$\mathcal{K} : C([0, T_1); D(A^{\alpha})) \rightarrow C([0, T_1); D(A^{\alpha})), \quad \forall T_1 > 0 \text{ and for } \alpha \in [0,1].$$

Also

$$\|\mathcal{K}v\|_{C([0,T_1);D(A^\alpha))} \leq \left\{ \sup_{t\in(0,T_1)} \int_0^t \frac{C_4}{(t-\tau)^\delta} \sup_{\tau\in(0,t)} \|L_u(\tau)\|_{\mathcal{L}(D(A^\alpha);X)} d\tau \right\} \|v\|_{C([0,T_1);D(A^\alpha))}$$

with $X = D(A^{\alpha-1/2})$ for $\alpha \in [0,1/2]$ and $X = E_{2\alpha-1}$ for $\alpha = 1/2$ or 1. That is

$$\|\mathcal{K}\|_{\mathcal{L}(C([0,T_1);D(A^\alpha));C([0,T_1);D(A^\alpha)))} \leq g_1(T_1)$$

with $g_1(T_1) \to 0$ as $T_1 \to 0$. This implies that for small T_1, $\mathcal{K}$ is a contraction in $C([0,T_1); D(A^\alpha))$.

Now note that for $v \in \mathcal{B}$,

$$\|[\mathcal{K}v](t+h) - [\mathcal{K}v](t)\|_{D(A^\alpha)} \leq C_5 \left\{ \int_0^t \left(\frac{1}{(t-\tau)^\delta} - \frac{1}{(t+h-\tau)^\delta} \right) \frac{1}{\tau^\gamma} d\tau \right.$$

$$\left. + \int_t^{t+h} \frac{1}{(t+h-\tau)^\delta} \frac{1}{\tau^\gamma} d\tau \right\} \|v\|_{\mathcal{B}}.$$

Here if $\alpha \in [0,1/2], \beta \in]-1/2,1/2]$ then $\gamma = \alpha - \beta$ and if $\alpha = 1/2$ or 1 and $v_0 \in E_{2\alpha-1}$ then $\gamma = (1+2\alpha)/4$. Since the right hand side for $t \in]0,T]$ tends to zero as $h \to 0$ we can deduce that $t \to [\mathcal{K}v](t)$ is continuous in $D(A^\alpha), \alpha \in [0,1]$ and $t \in]0,T]$ if $v \in \mathcal{B}$. We also obtain

$$\|\mathcal{K}v\|_{\mathcal{B}} \leq C_6 \left\{ \sup_{t\in(0,T_1)} t^\gamma \int_0^t \frac{1}{\tau^\gamma(t-\tau)^\delta} \sup_{\tau\in(0,t)} \|L_u(\tau)\|_{\mathcal{L}(D(A^\alpha);X)} d\tau \right\} \|v\|_{\mathcal{B}}.$$

That is

$$\|\mathcal{K}\|_{\mathcal{L}(\mathcal{B};\mathcal{B})} \leq g_2(T_1) \text{ with } g_2(T_1) \to 0 \text{ as } T_1 \to 0.$$

This implies that for small T_1, $\mathcal{K}$ is a contraction in $\mathcal{B}$.

♣

Lemma 4.7 *For $v \in C([0,T_1); D(A^\alpha)), \alpha \in [0,1]$ the time derivative*

$$[\mathcal{K}v]'(\cdot) \in C((0,T_1); D(A^{\alpha-1})).$$

Moreover when $v \in \mathcal{B}$ the map $t \to [\mathcal{K}v]'(t)$ is continuous from $]0,T_1)$ in to $D(A^{\alpha-1})$ for $\alpha \in [0,1]$.

Proof: Let $v \in C([0, T_1); D(A^\alpha))$ and consider,

$$\frac{1}{h}([\mathcal{K}v](t+h) - [\mathcal{K}v](t)) = -\frac{(S(h) - I)}{h} \int_0^t S(t - \tau) L_u(\tau) v(\tau) d\tau$$

$$-\frac{1}{h} \int_t^{t+h} S(t + h - \tau) L_u(\tau) v(\tau) d\tau.$$

Taking the limit $h \to 0$ to get the right derivative,

$$[\mathcal{K}v]'_+(t) = -A[\mathcal{K}v](t) - L_u(t) v(t).$$

Using the properties of the operators A and L_u we conclude that $[\mathcal{K}v]'_+(\cdot) \in C((0, T_1); D(A^{\alpha-1}))$.
Let us define $W(t)$ by

$$W(t) = [\mathcal{K}v](0) + \int_0^t [\mathcal{K}v]'_+(\tau) d\tau \tag{4.13}$$

and note that $W'(\cdot) \in C((0, T_1); D(A^{\alpha-1}))$. Now let $\Theta(t) = W(t) - [\mathcal{K}v](t)$. Then $\Theta'_+(t) = 0$
and $\Theta(0) = 0$. Let us define a scalar function $\theta(t)$ using the inner product in $D(A^{\alpha-1})$ as

$$\theta(t) = (\Theta(t), u)_{D(A^{\alpha-1})}, u \in D(A^{\alpha-1}).$$

We then get $\theta(0) = 0$ and $\theta'_+(t) = 0$ for $t \in (0, T_1)$. Note that these properties of θ
imply $\theta(t) \leq 0$ for $t \in (0, T_1)$. If not there exists $t_1 \in (0, T_1)$ for which $\theta(t_1) > 0$. Let
$t_0 = \inf\{t, \theta(t) > 0\}$. By the continuity of $\theta(t)$, $\theta(t_0) = 0$ and there exists a sequence
$\{t_n\} \geq t_0$ such that $t_n \to t_0$ and $\theta(t_n) > 0$. Thus

$$\theta'_+(t_0) = \lim_{t_n \to t_0} \frac{\theta(t_n) - \theta(t_0)}{t_n - t_0} > 0 \tag{4.14}$$

which is a contradiction since $\theta'_+(t) = 0$. By a similar arguement we can show that $\theta(t) \geq 0$
for $t \in (0, T_1)$. Thus $\theta(t) = 0, \forall t \in (0, T_1)$. This means, $\forall u \in D(A^{\alpha-1})$ and $\forall t \in (0, T_1)$,
$(\Theta(t), u)_{D(A^{\alpha-1})} = 0$ which implies $\Theta(t) = 0$ in $D(A^{\alpha-1})$, $\forall t \in (0, T_1)$. Hence $[\mathcal{K}v](t) = W(t)$
and is continuously differentiable in $D(A^{\alpha-1})$.

We now consider the case when $v \in \mathcal{B}$. In this case the map $t \to v(t)$ is continuous from
$]0, T_1]$ in to $D(A^\alpha), \alpha \in [0, 1]$. This implies that the maps $t \to L_u(t) v(t)$ and $t \to A[\mathcal{K}v](t)$
are continuous from $]0, T_1]$ in to $D(A^{\alpha-1})$. We thus conclude using the arguements of the
first case that $[\mathcal{K}v](t)$ is differentiable in $D(A^{\alpha-1})$ for $t \in]0, T_1)$.

♣

Let us now characterize the evolution operator associated to the linear differential equation in problem 4.1. We write the representation (4.12) in problem 4.2 as

$$[(I - \mathcal{K}v)](t) = S(t)v_0.$$

Then for small enough T_1, the operator $I - \mathcal{K}$ is invertible in $C([0, T_1); D(A^\alpha))$ and in $\mathcal{B}$ since $\mathcal{K}$ is a contraction in these spaces. We obtain a convergent series(in these Banach spaces) as

$$v(\cdot) = \sum_{n=0}^{\infty} [\mathcal{K}^n S](\cdot)v_0. \tag{4.15}$$

We write this as $v(t) = Z(t,0)v_0$ and call $Z(t,0)$ the evolution operator. Note that the convergence of this series ensures the uniqueness of the solution to problem 4.2 (and hence to the problem 4.1).

Let us now study the evolution for $t \geq \tau$ by prescribing the initial data at $t = \tau$ in problems 4.1 and 4.2. The evolution operator obtained (as a series) in this manner is denoted $Z(t,\tau)$ with $t - \tau \leq T_1$. Here T_1 is taken small enough to ensure the convergence of the series. Let us consider for $0 \leq \eta \leq \tau \leq t$, $v(t) = Z(t,\tau)v(\tau)$ with $v(\tau) = Z(\tau,\eta)v_0$. That is

$$v(t) = Z(t,\tau)Z(\tau,\eta)v_0.$$

Due to the uniqueness of the solution we have

$$v(t) = Z(t,\eta)v_0 = Z(t,\tau)Z(\tau,\eta)v_0.$$

That is

$$Z(t,\eta) = Z(t,\tau)Z(\tau,\eta), 0 \leq \eta \leq \tau \leq t.$$

Iterating this kind of arguements we can extend the definition of $Z(t_2, t_1)$ to $t_2 - t_1 \in [0, \infty)$.

Let us now provide a formal proof of the above arguement.

Lemma 4.8 *The evolution operator $Z(\cdot, \cdot)$ satisfies*

$$Z(t,\tau) = Z(t,\eta)Z(\eta,\tau), 0 \leq \tau \leq \eta \leq t < \infty \tag{4.16}$$

Proof

Let us define a linear operator $\mathcal{K}_\tau$ such that

$$[\mathcal{K}_\tau v](t) = 0 \cdots 0 \le t \le \tau \le T_2$$

$$[\mathcal{K}_\tau v](t) = -\int_\tau^t S(t-\eta)L_u(\eta)v(\eta)d\eta \cdots 0 \le \tau \le t \le T_2.$$

Then for sufficiently small T_2, $\mathcal{K}_\tau$ is a contraction in the Banach spaces $C([0, T_2); D(A^\alpha)), \alpha \in [0, 1]$ and $\mathcal{B}$. Here $\mathcal{B}$ has to be modified with an appropriate time shift. Namely

$$\mathcal{B} = \{t \to v(t); v(t) \text{ continuous in } D(A^\alpha), \alpha \in [0, 1] \text{ for } t \in (\tau, T_2)$$

and $(t-\tau)^\gamma v(t)$ bounded in $D(A^\alpha)\}$ with norm $\|v\|_{\mathcal{B}} = \sup_{t \in (\tau, T_2)} \|(t-\tau)^\gamma v(t)\|_{D(A^\alpha)}\}$.

We then get the following series representation for $v(t)$ that converges in the above Banach spaces for sufficiently small T_2:

$$v(\cdot) = Z(\cdot, \tau)v_0 = S(\cdot - \tau)v_0 + \sum_{n \ge 1}[\mathcal{K}_\tau^n S](\cdot - \tau)v_0. \tag{4.17}$$

This defines $Z(t, \tau)$ for $0 \le \tau \le t \le T_2$. Now using the uniqueness result we can write

$$Z(t, \tau) = Z(t, \eta)Z(\eta, \tau), 0 \le \tau \le \eta \le t \le 2T_2.$$

Repeating this kind of representation we can define $Z(t, \tau)$ for $0 \le \tau \le t < \infty$.

♣

The next series of lemmas provide useful regularity properties of $Z(\cdot, \cdot)$.

Lemma 4.9 *For $0 \le \tau < t$ the evolution operator satisfies the following estimates:*

(i) For $\alpha \in [0, 1]$

$$\|Z(t, \tau)\|_{\mathcal{L}(D(A^\alpha); D(A^\alpha))} \le C_8 e^{\sigma_1(t-\tau)}.$$

(ii) For $\alpha_1 \in] -1/2, 1/2], \alpha_2 \in [0, 1/2]$ with $\alpha_1 \le \alpha_2$

$$\|Z(t, \tau)\|_{\mathcal{L}(D(A^{\alpha_1}); D(A^{\alpha_2}))} \le C_9\{1 + \frac{1}{(t-\tau)^{\alpha_2 - \alpha_1}}\}e^{\sigma_2(t-\tau)}.$$

(iii) For $\alpha = 1/2$ or 1 then,

$$\|Z(t,\tau)\|_{\mathcal{L}(E_{2\alpha-1};D(A^\alpha))} \leq C_{10}\{1 + \frac{1}{(t-\tau)^{\alpha/2+1/4}}\}e^{\sigma_3(t-\tau)}.$$

Here $\sigma_1, \sigma_2, \sigma_3 \geq 0$ and $C_8, C_9, C_{10} > 0$.

Proof

Let us first consider the case of $0 \leq \tau < t \leq T_2$. Estimating the series representation we get

$$\|Z(\cdot,\tau)\|_{D(A^\alpha)} \leq \|S(\cdot-\tau)v_0\|_{D(A^\alpha)} + \sum_{n\geq 1} \|[\mathcal{K}^n_\tau S](\cdot-\tau)v_0\|_{D(A^\alpha)}.$$

First take $\alpha_2 = \alpha_1 = \alpha$. Then

$$\|Z(t,\tau)v_0\|_{D(A^\alpha)} \leq \|v_0\|_{D(A^\alpha)} + \sum_{n\geq 1} \|[\mathcal{K}^n_\tau S](t-\tau)v_0\|_{D(A^\alpha)}.$$

Now note that for $u \in C(0,T_2;D(A^\alpha))$, $\alpha \in [0,1]$ we have

$$\|[\mathcal{K}^n_\tau u](t)\|_{D(A^\alpha)} \leq g_1(t,\tau)r_1^{n-1}\|u\|_{C(0,T_1;D(A^\alpha))}$$

where

$$\|\mathcal{K}_\tau\|_{\mathcal{L}(C(0,T_1;D(A^\alpha));C(0,T_1;D(A^\alpha)))} \leq r_1 < 1$$

and $g_1(t,\tau) \to 0$ as $t \to \tau$. Now set $u(t) = S(t-\tau)v_0$ to get

$$\|Z(t,\tau)\|_{\mathcal{L}(D(A^\alpha);D(A^\alpha))} \leq [1 + \frac{g_1(t,\tau)}{1-r_1}]$$

$$\leq \text{const.}e^{\sigma_1(t-\tau)} \text{ for } 0 \leq \tau \leq t \leq T_2.$$

For larger values we iterate this result using the lemma 4.8 to get

$$\|Z(t,\tau)\|_{\mathcal{L}(D(A^\alpha);D(A^\alpha))} \leq C_8 e^{\sigma_1(t-\tau)} \quad 0 \leq \tau \leq t, \alpha \in [0,1].$$

Recall that for $t > \tau \geq 0$ and $\alpha = 1/2$ or 1,

$$\|S(t-\tau)v_0\|_{D(A^\alpha)} \leq \frac{\text{const.}}{(t-\tau)^{\alpha/2+1/4}}\|v_0\|_{E_{2\alpha-1}}$$

and

$$\|S(t-\tau)v_0\|_{D(A^{\alpha_2})} \leq \frac{\text{const.}}{(t-\tau)^{\alpha_2-\alpha_1}}\|v_0\|_{D(A^{\alpha_1})}$$

for $\alpha_1 \in]-1/2, 1/2]$ and $\alpha_2 \in [0, 1/2]$.

Now for $u \in \mathcal{B}$

$$\|[\mathcal{K}_\tau^n u](t)\|_{D(A^\alpha)} \leq g_2(t,\tau) r_2^{n-1} \|u\|_{\mathcal{B}}, \alpha \in [0,1]$$

where

$$\|\mathcal{K}_\tau\|_{\mathcal{L}(\mathcal{B};\mathcal{B})} \leq r_2 < 1.$$

Now set $u(t) = S(t-\tau)v_0$ and note that for $v_0 \in E_{2\alpha-1}, \alpha = 1/2$ or 1,

$$\|S(\cdot - \tau)v_0\|_{\mathcal{B}} \leq \text{const.}\|v_0\|_{E_{2\alpha-1}}$$

and for $v_0 \in D(A^{\alpha_1}), \alpha_1 \in [-1/2, 1/2]$

$$\|S(\cdot - \tau)v_0\|_{\mathcal{B}} \leq \text{const.}\|v_0\|_{D(A^{\alpha_1})}.$$

Hence we get,

$$\|Z(t,\tau)\|_{\mathcal{L}(E_{2\alpha-1};D(A^\alpha))} \leq \text{const.}\{\frac{1}{(t-\tau)^{\frac{\alpha}{2}+\frac{1}{4}}} + \frac{g_2(t,\tau)}{1-r_2}\},$$

and

$$\|Z(t,\tau)\|_{\mathcal{L}(D(A^{\alpha_1});D(A^{\alpha_2}))} \leq \text{const.}\{\frac{1}{(t-\tau)^{\alpha_2-\alpha_1}} + \frac{g_3(t,\tau)}{1-r_2}\}.$$

For larger values of t we will use

$$\|Z(t,\tau)\|_{\mathcal{L}(E_{2\alpha-1};D(A^\alpha))} \leq \|Z(t,\eta)\|_{\mathcal{L}(D(A^\alpha);D(A^\alpha))} \|Z(\eta,\tau)\|_{\mathcal{L}(E_{2\alpha-1};D(A^\alpha))}$$

and

$$\|Z(t,\tau)\|_{\mathcal{L}(D(A^{\alpha_1});D(A^{\alpha_2}))} \leq \|Z(t,\eta)\|_{\mathcal{L}(D(A^{\alpha_2});D(A^{\alpha_2}))} \|Z(\eta,\tau)\|_{\mathcal{L}(D(A^{\alpha_1});D(A^{\alpha_2}))}.$$

♣

Lemma 4.10 *For $\alpha \in [0,1]$ and $0 \leq t_1 \leq t_2 < \infty$ we have as $t_2 \to t_1^+, Z(t_2, t_1) \to I$ strongly in $\mathcal{L}(D(A^\alpha); D(A^\alpha))$. For $t > \tau$ the map $t \to Z(t,\tau)$ is continuous in the uniform operator topology of $\mathcal{L}(D(A^\alpha); D(A^\alpha))$ and for $\tau < t$ the map $\tau \to Z(t,\tau)$ is continuous in the uniform operator topology of $\mathcal{L}(D(A^\alpha); D(A^\alpha))$*

Proof

To establish the strong continuity we will consider $v_0 \in D(A^\alpha), \alpha \in [0,1]$,

$$Z(\tau + h, \tau)v_0 - v_0 = S(h)v_0 - v_0 + \sum_{n \geq 1}[\mathcal{K}_\tau^n S](\tau + h - \tau)v_0.$$

Noting the fact that $S(t)$ is strongly continuous in $D(A^\alpha)$ at the origin,

$$\|Z(\tau + h, \tau)v_0 - v_0\|_{D(A^\alpha)} \leq \|S(h)v_0 - v_0\|_{D(A^\alpha)} + g_0(\tau; h)\|v_0\|_{D(A^\alpha)}$$

$$\to 0 \text{ as } h \to 0, \forall v_0 \in D(A^\alpha).$$

We will now establish the continuity in the uniform operator topology. Note that for any $u \in C(0, T_2; D(A^\alpha)), \tau < t$

$$\|[\mathcal{K}_\tau u](t + h) - [\mathcal{K}_\tau u](t)\|_{D(A^\alpha)} \leq g_1(t, \tau, h)\|u\|_{C(0,T;D(A^\alpha))}$$

where $g_1(t, \tau, h) \to 0$ as $h \to 0$. Also

$$\|[\mathcal{K}_\tau^n u](t + h) - [\mathcal{K}_\tau^n u](t)\|_{D(A^\alpha)} \leq g_1(t, \tau, h)r_1^{n-1}\|u\|_{C(0,T_2;D(A^\alpha))}.$$

Here $r_1 < 1$ for sufficiently small T_2. Hence setting $S(t - \tau)v_0 = u(t)$ we get

$$\|Z(t + h, \tau)v_0 - Z(t, \tau)v_0\|_{D(A^\alpha)} \leq \|S(t + h - \tau) - S(t - \tau)\|_{\mathcal{L}(D(A^\alpha);D(A^\alpha))}\|v_0\|_{D(A^\alpha)}$$

$$+ \frac{g_1(t, \tau, h)}{1 - r_1}\|v_0\|_{D(A^\alpha)}.$$

Since $S(t), t > 0$ is analytic and hence continuous in the uniform operator topology,

$$\|Z(t + h, \tau) - Z(t, \tau)\|_{\mathcal{L}(D(A^\alpha);D(A^\alpha))} \to 0 \text{ as } h \to 0.$$

This proves the continuity of the map $t \to Z(t, \tau)$ in the uniform operator topology for $\tau < t$. Let us now establish the continuity of the map $\tau \to Z(t, \tau)$ for $\tau < t$. Note that from the analyticity of the Stokes semigroup, we have for $t > \tau$

$$\|S(t - \tau - h) - S(t - \tau)\|_{\mathcal{L}(D(A^\alpha);D(A^\alpha))} \to 0 \text{ as } h \to 0.$$

Now note that for $u_\tau \in C([0, T_2); D(A^\alpha))$

$$\|[\mathcal{K}_{\tau+h} u_{\tau+h}](\cdot) - [\mathcal{K}_\tau u_\tau](\cdot)\|_{C([0,T);D(A^\alpha))} \leq g_3(T_2, \tau, h)\|u_{\tau+h}(\cdot) - u_\tau(\cdot)\|_{C([0,T_2);D(A^\alpha))}$$

$$+ g_4(T_2, \tau, h)\|u_\tau(\cdot)\|_{C([0,T_2);D(A^\alpha))}$$

with $g_4(T_2, \tau, h) \to 0$ as $h \to 0$.

Hence by substituting for $u_\tau(\cdot)$ successively the elements

$$S(\cdot - \tau)v_0, [\mathcal{K}_\tau S](\cdot - \tau)v_0, \cdots, [\mathcal{K}_\tau^n S](\cdot - \tau)v_0, \cdots \text{ etc}$$

we conclude that the maps $\tau \to [\mathcal{K}_\tau^n S](\cdot - \tau)$ are continuous from

$$[0, T_2] \text{ in to } \mathcal{L}(D(A^\alpha); C([0, T_2); D(A^\alpha))).$$

That is the map $\tau \to Z(t, \tau)$ is continuous in $\mathcal{L}(D(A^\alpha); C([0, T_2); D(A^\alpha)))$ for $\alpha \in [0, 1]$.

♣

Lemma 4.11 *For $t > \tau$ and $\alpha = 1/2$ or 1 the map $t \to Z(t, \tau)$ is continuous in the uniform operator topology of $\mathcal{L}(E_{2\alpha-1}; D(A^\alpha))$ and for $\tau < t$ the map $\tau \to Z(t, \tau)$ is continuous in the uniform operator topology of $\mathcal{L}(E_{2\alpha-1}; D(A^\alpha))$.*

Proof: Note that for any $u \in \mathcal{B}, \tau < t$

$$\|[\mathcal{K}_\tau u](t + h) - [\mathcal{K}_\tau u](t)\|_{D(A^\alpha)} \leq g_2(t, \tau, h)\|u\|_\mathcal{B}$$

where $g_2(t, \tau, h) \to 0$ as $h \to 0$. Also

$$\|[\mathcal{K}_\tau^n u](t + h) - [\mathcal{K}_\tau^n u](t)\|_{D(A^\alpha)} \leq g_2(t, \tau, h)r_2^{n-1}\|u\|_\mathcal{B}.$$

Here $r_2 < 1$ for sufficiently small T. Hence setting $S(t - \tau)v_0 = u(t)$ we get

$$\|Z(t + h, \tau)v_0 - Z(t, \tau)v_0\|_{D(A^\alpha)} \leq \|S(t + h - \tau) - S(t - \tau)\|_{\mathcal{L}(E_{2\alpha-1};D(A^\alpha))}\|v_0\|_{E_{2\alpha-1}}$$

$$+ \frac{g_2(t, \tau, h)}{1 - r_2}\|v_0\|_{E_{2\alpha-1}}.$$

Hence

$$\|Z(t+h,\tau) - Z(t,\tau)\|_{\mathcal{L}(E_{2\alpha-1};D(A^\alpha))} \to 0 \text{ as } h \to 0.$$

Let us now establish the continuity of the map $\tau \to Z(t,\tau)$. First note that for $\tau < t$, using the lemma 4.2

$$\|S(\cdot - \tau - h)v_0 - S(\cdot - \tau)v_0\|_{\mathcal{B}} \leq \text{const.}g_6(h)\|v_0\|_{E_{2\alpha-1}}$$

$$\to 0 \text{ as } h \to 0.$$

Let us consider for $u_\tau \in \mathcal{B}$

$$\|[\mathcal{K}_{\tau+h}u_{\tau+h}](\cdot) - [\mathcal{K}_\tau u_\tau](\cdot)\|_{\mathcal{B}} \leq g_3(T_2,\tau,h)\|u_{\tau+h}(\cdot) - u_\tau(\cdot)\|_{\mathcal{B}} + g_4(T_2,\tau,h)\|u_\tau(\cdot)\|_{\mathcal{B}}$$

with $g_4(T_2,\tau,h) \to 0$ as $h \to 0$. Hence by substituting for $u_\tau(\cdot)$ successively the elements $S(\cdot-\tau)v_0, [\mathcal{K}_\tau S](\cdot-\tau)v_0, \cdots, [\mathcal{K}_\tau^n S](\cdot-\tau)v_0, \cdots$ etc we conclude that the maps $\tau \to [\mathcal{K}_\tau^n S](\cdot-\tau)$ are continuous from $[0,T_2]$ in to $\mathcal{L}(E_{2\alpha-1};\mathcal{B})$. That is the map $\tau \to Z(t,\tau)$ is continuous in $\mathcal{L}(E_{2\alpha-1};\mathcal{B})$.

♣

By a similar arguement we can prove

Lemma 4.12 *Let $\alpha_1 \in [-1/2, 1/2]$ and $\alpha_2 \in [0, 1/2]$ with $\alpha_1 \leq \alpha_2$. Then for $t > \tau$ the map $t \to Z(t,\tau)$ is continuous in the uniform operator topology of $\mathcal{L}(D(A^{\alpha_1}); D(A^{\alpha_2}))$ and for $\tau < t$ the map $\tau \to Z(t,\tau)$ is continuous in the uniform operator topology of $\mathcal{L}(D(A^{\alpha_1}); D(A^{\alpha_2}))$.*

Let us now establish the compactness of the evolution operator.

Theorem 4.8 *For $0 \leq \tau < t$ the operator*

$$Z(t,\tau) \in \mathcal{L}(D(A^{\alpha_0}); D(A^{\alpha_0})) \cap \mathcal{L}(D(A^{\alpha_1}); D(A^{\alpha_2})) \cap \mathcal{L}(E_{2\alpha-1}; D(A^\alpha))$$

is compact with $\alpha_0 \in [0,1], \alpha_1 \in]-1/2, 1/2], \alpha_2 \in [0,1/2], \alpha_2 \geq \alpha_1$ and $\alpha = 1/2$ or 1.

Proof: Let us define a linear operator $G_\epsilon(t)$ such that

$$G_\epsilon(t)v_0 = \int_\tau^{t-\epsilon} S(t-\eta)L_u(\eta)w(\eta)d\eta \text{ for } t > \epsilon + \tau$$

$$G_\epsilon(t)v_0 = 0 \text{ for } t \leq \epsilon + \tau, \epsilon > 0.$$

Here the map, $v_0 \to w(t)$ is continuous linear in $D(A^{\alpha_0})$. Let us define the Riemann sums:

$$R_N(t)v_0 = \sum_{j=1}^{N} S(t-\eta_j)L_u(\eta_j)w(\eta_j)(\eta_{j+1} - \eta_j), \text{ with}$$

$$\tau = \eta_1 < \eta_2 < \cdot < \eta_{N+1} = t - \epsilon.$$

Observe that $G_\epsilon(t)$ is the limit of $R_N \in \mathcal{L}(D(A^{\alpha_0}); D(A^{\alpha_0}))$ in the uniform operator topology. Note that the map $v_0 \to w(\eta_j)$ is continuous linear in $D(A^{\alpha_0})$ and

$$L_u(\eta_j) \in \mathcal{L}(D(A^\alpha); E_{2\alpha-1}) \cap \mathcal{L}(D(A^{\alpha_2}); D(A^{\alpha_2 - 1/2})).$$

Also as established in an earlier section the semigroup

$$S(\zeta) \in \mathcal{L}(E_{2\alpha-1}; D(A^\alpha)) \cap \mathcal{L}(D(A^{\alpha_2 - 1/2}); D(A^{\alpha_2}))$$

is compact for $\zeta > 0$. Since $t - \eta_j > 0, \forall j$ we have the map $v_0 \to S(t - \eta_j)L_u(\eta_j)w(\eta_j)$ compact in $D(A^{\alpha_0}), \forall j$. Thus $R_N(t) \in \mathcal{L}(D(A^{\alpha_0}); D(A^{\alpha_0}))$ is compact for $t > \tau + \epsilon$ and for each N. This implies that the map $G_\epsilon(t) \in \mathcal{L}(D(A^{\alpha_0}); D(A^{\alpha_0}))$ is compact $\forall \epsilon > 0$ and $\forall t > \tau + \epsilon$. Taking the limit $\epsilon \to 0$ we deduce that $G_0(t)$ is compact in $D(A^{\alpha_0})$. Now note that the evolution operator $Z(t, \tau)$ can be written as (for small t)

$$Z(\cdot, \tau)v_0 = S(\cdot - \tau)v_0 + \sum_{n \geq 1}[\mathcal{K}_\tau^n S](\cdot - \tau)v_0.$$

Now setting succesively the terms $S(t - \tau)v_0, [\mathcal{K}_\tau S](t - \tau)v_0, \cdots, [\mathcal{K}_\tau^n S](t - \tau)v_0, \cdots$ for $w(t)$ and using the compactness of $G_0(t)$ in $D(A^{\alpha_0})$ we deduce that each term in the series representation for $Z(t, \tau)$ is compact in $D(A^{\alpha_0})$. This in combination with the fact that for all $t \leq T_2$, this series converges uniformly in $D(A^{\alpha_0})$ we deduce that the map $Z(t, \tau) \in \mathcal{L}(D(A^{\alpha_0}); D(A^{\alpha_0}))$ is compact for $t > \tau$.

We can now deduce the compactness of the operator $Z(t,\eta) \in \mathcal{L}(D(A^{\alpha_1}); D(A^{\alpha_2}))$ by writing $Z(t,\eta) = Z(t,\tau)Z(\tau,\eta), 0 \leq \eta < \tau < t$ and noting the boundedness of $Z(\tau,\eta) \in \mathcal{L}(D(A^{\alpha_1}); D(A^{\alpha_2}))$ and the compactness of $Z(t,\tau) \in \mathcal{L}(D(A^{\alpha_2}); D(A^{\alpha_2}))$. Similarly the compactness of $Z(t,\eta) \in \mathcal{L}(E_{2\alpha-1}; D(A^{\alpha}))$ can be deduced from the boundedness of $Z(\tau,\eta) \in \mathcal{L}(E_{2\alpha-1}; D(A^{\alpha}))$ and the compactness of $Z(t,\tau) \in \mathcal{L}(D(A^{\alpha}); D(A^{\alpha}))$.

♣

From the results obtained so far we can deduce the following theorem for the spectrum of $Z(t_2,t_1)$.

Lemma 4.13 *The spectrum of $Z(t_2,t_1), t_2 > t_1 \geq 0$ is*

(i) discrete with finite multiplicity and accumulation possible only at the origin.

(ii) same in $D(A^{\alpha}), \forall \alpha \in [0,1]$.

Proof

Note that $\forall \alpha \in [0,1]$ the evolution operator $Z(t_2,t_1) : D(A^{\alpha}) \to D(A)$ is compact and this gives (i). Moreover the eigenvectors are in $D(A)$ and thus the spectrum is same in $D(A^{\alpha}), \forall \alpha \in [0,1]$.

♣

Let us now specialize our study to the case where the **basic field U is T-periodic in time**.

Lemma 4.14 *Let the basic field be T-periodic in time. Then*

(i) $Z(nT,0) = Z(T,0)^n$, $\forall n \geq 1$,

(ii) The spectrum of $Z(T + t_0, t_0)$ is independent of $t_0 \geq 0$.

Proof:

First note that due to the periodicity of the basic field U we have $L_u(t + T) = L_u(t), t \geq 0$. This implies

$$\int_{t_1+T}^{t_2+T} S(t_2 + T - \eta)L_u(\eta)S(\eta - t_1 - T)d\eta = \int_{t_1}^{t_2} S(t_2 - \zeta)L_u(\zeta)S(\zeta - t_1)d\zeta.$$

Hence we conclude that

$$Z(t_2 + T, t_1 + T) = Z(t_2, t_1), t_2 \geq t_1 \geq 0. \tag{4.18}$$

We have from equation (4.16)

$$Z(t_2 + T, t_1) = Z(t_2 + T, t_1 + T)Z(t_1 + T, t_1)$$

valid for any time dependent basic field. Combining (4.16) and (4.18) gives

$$Z(t_2 + T, t_1) = Z(t_2, t_1)Z(t_1 + T, t_1).$$

Set $t_1 = 0$ and $t_2 = (n-1)T, n \geq 1$ to get

$$Z(nT, 0) = Z((n-1)T, 0)Z(T, 0).$$

From this we can get successively,

$$Z(nT, 0) = [Z(T, 0)]^n, n \geq 1.$$

To establish part(ii) let σ belongs to the spectrum of $Z(T + t_0, t_0)$. We will show that σ will also belong to the spectrum of $Z(T, 0)$. We have $Z(T + t_0, t_0)\phi = \sigma\phi$ where ϕ is the eigenvector. Let us define $\psi = Z(nT, t_0)\phi$ with $nT \geq t_0$. Then we have due to (4.18),

$$Z(T, 0)\psi = Z(T + nT, nT)Z(nT, t_0)\phi.$$

Now using (4.16) to the right hand side we get,

$$Z(T, 0)\psi = Z(T + nT, t_0)\phi$$

$$= Z(T + nT, T + t_0)Z(T + t_0, t_0)\phi$$

$$= \sigma Z(nT, t_0)\phi = \sigma\psi.$$

On the other hand let μ belongs to the spectrum of $Z(T, 0)$ and let v be the corresponding eigenvector. Let us set $w = Z(nT + t_0, 0)v, t_0 \geq 0$. Using (4.18) we get,

$$Z(T + t_0, t_0)w = Z((n+1)T + t_0, nT + t_0)Z(nT + t_0, 0)v$$

$$= Z((n+1)T + t_0, 0)\boldsymbol{v}$$

$$= Z((n+1)T + t_0, T)Z(T, 0)\boldsymbol{v}$$

$$= \mu Z(nT + t_0, 0)\boldsymbol{v} = \mu\boldsymbol{w}$$

Thus the spectrum of $Z(T + t_0, t_0)$ does not depend on $t_0 \geq 0$.

$$\clubsuit$$

We will call the operator $Z(T, 0)$ the **Monodromy operator**. For a study of such operators in general evolution problems see [16].

Remark:

We note here that a stationary basic flow can be regarded as a time periodic basic flow with arbitrary period. In this case the evolution operator $Z(t, \tau)$ coincides with the operator $Z(t - \tau)$ defined in section 4.2.

Chapter 5

The Nonlinear Hydrodynamic Semigroup

Let us now characterize the nonlinear semigroup associated with the Navier-Stokes system 2.2 of Chapter 2. Our goal is to define the discrete time-T map which relates the velocity field at time-T (which is a point in an appropriate function space) to the velocity field at time zero (which is again a point in this space). In addition we will prove that this dependence is Frechet analytic. This last result will be used in chapter 7 to establish the analyticity of the invariant manifolds. We will treat the case of bounded domains first. Unbounded domains will be discussed in section 5.1.

Let us now consider the evolution form of the full Navier-Stokes system 2.2

Problem 5.1 *Find* $v \in C([0, T^*); D(A^\alpha)) \cap C^1(0, T^*; D(A^{\alpha-1}))$ *such that*

$$\frac{dv}{dt} + Av + L_u(t)v + B(v, v) = 0, t \in (0, T^*) \tag{5.1}$$

$$v(0) = v_0 \in D(A^\alpha), \alpha \in [1/2, 1].$$

Here the bilinear operator $B(\cdot, \cdot)$ is defined using the Riesz representation theorem as

$$< B(v, u), w >_{V' \times V} = b(v, u, w), \forall w \in V.$$

The evolution problem 5.1 can be derived by simply applying the projection operator P_H on to the system 2.2. The maximal time T^* will depend on the norm of the initial data and will be determined by the unique solvability theorem.

Let us consider the following integral representation:

Problem 5.2 *Find* $v \in C([0, T^*); D(A^\alpha))$ *such that*

$$v(t) = Z(t, 0)v_0 - \int_0^t Z(t, \eta)B(v(\eta), v(\eta))d\eta \tag{5.2}$$

$$v(0) = v_0 \in D(A^\alpha), \alpha \in [1/2, 1].$$

Theorem 5.1 *Problem 5.2 resolves problem 5.1.*

Proof: First note that the representation (5.2) in problem 5.2 formally satisfies the evolution form (5.1) in problem 5.1. From the regularity properties of the evolution operator we have,

$$Z(t, 0)v_0 \in C([0, \infty); D(A^\alpha)) \cap C^1(0, \infty; D(A^{\alpha-1})). \tag{5.3}$$

Let us set

$$Y(t) = \int_0^t Z(t, \eta)B(v(\eta), v(\eta))d\eta. \tag{5.4}$$

We only need to show that $Y(\cdot) \in C([0, T^*); D(A^\alpha)) \cap C^1(0, T^*; D(A^{\alpha-1}))$.

Let us first show that $Y(t)$ is bounded and continuous in $D(A^\alpha)$. Note that for $v \in C([0, T^*); D(A^\alpha))$ we have $B(v, v) \in C([0, T^*); X)$ with $X = E_1$ for $\alpha = 1$ and $X = D(A^{-1/4})$ for $\alpha = 1/2$. This follows from the following observation. Firstly from the general estimate for the trilinear form given in chapter 3, we have for

$$|b(v, v, w)| \leq \text{ const. } \|v\|_{H^1(\Omega)}\|v\|_{H^1(\Omega)}\|w\|_{H^{1/2}(\Omega)}, \forall v \in V \text{ and } \forall w \in H^{1/2}(\Omega).$$

Hence by the Riesz representation theorem we can write

$$b(v, v, w) = <B(v, v), w>_{D(A^{-1/4}) \times D(A^{1/4})}, \forall w \in D(A^{1/4}).$$

This gives us the result:

$$B(\cdot, \cdot) : D(A^{1/2}) \times D(A^{1/2}) \to D(A^{-1/4}).$$

90

It is possible to obtain such a result for unbounded domains by far more involved methods (see section 5.1). We now consider the estimate,

$$|b(v, v, w)| \leq \text{const.} \, \|v\|_{H^2(\Omega)} \|v\|_{H^2(\Omega)} \|w\|_{L^2(\Omega)} \forall v \in D(A) \text{ and } \forall w \in H$$

This gives, as before $B(v, v) \in H$. We then note that $H^2(\Omega)$ is an algebra for $\Omega \subset R^3$ and hence $v_i v_j \in H^2(\Omega)$ for $v \in D(A)$. Now, since $B(v, v) = P_H[D_i(v_i v)]$ and $P_H : H^1(\Omega) \to H^1(\Omega) \cap H = E_1$ we get $B(v, v) \in E_1$.

Let us now estimate (5.4) as,

$$\|Y(t)\|_{D(A^\alpha)} \leq \int_0^t \|Z(t, \eta)\|_{\mathcal{L}(X; D(A^\alpha))} \|B(v(\eta), v(\eta))\|_X d\eta.$$

Now using the estimates for $B(\cdot, \cdot)$ and the estimates for the evolution operator obtained in the previous chapter we get,

$$\|Y\|_{C([0,T^*); D(A^\alpha))} \leq y_1(T^*) \|v\|^2_{C([0,T^*); D(A^\alpha))}. \tag{5.5}$$

Let us now consider,

$$Y(t + h) - Y(t) = \int_0^t [Z(t + h, \eta) - Z(t, \eta)] B(v(\eta), v(\eta)) d\eta$$

$$+ \int_t^{t+h} Z(t + h, \eta) B(v(\eta), v(\eta)) d\eta.$$

Thus

$$\|Y(t + h) - Y(t)\|_{D(A^\alpha)} \leq \left\{ \int_0^t \|Z(t + h, \eta) - Z(t, \eta)\|_{\mathcal{L}(X; D(A^\alpha))} d\eta \right.$$

$$\left. + \int_t^{t+h} \|Z(t + h, \eta)\|_{\mathcal{L}(X; D(A^\alpha))} d\eta \right\} \|B(v, v)\|_{C(0, T^*; X)}.$$

We get again,

$$\|Y(t + h) - Y(t)\|_{D(A^\alpha)} \leq y_2(t, h) \|v\|^2_{C([0,T^*); D(A^\alpha))}$$

with $y_2(t, h) \to 0$ as $h \to 0$. We hence conclude that $Y \in C([0, T^*); D(A^\alpha))$. Now

$$\frac{Y(t + h) - Y(t)}{h} = \left(\frac{Z(t + h, t) - Z(t, t)}{h} \right) \int_0^t Z(t, \eta) B(v(\eta), v(\eta)) d\eta$$

$$+ Z(t + h, t) \frac{1}{h} \int_t^{t+h} Z(t, \eta) B(v(\eta), v(\eta)) d\eta.$$

Taking the limit $h \to 0$ we get due to the continuity results of the evolution operator $Z(\cdot,\cdot)$ obtained in chapter 4,

$$Y'_+(t) = -(L_u(t) + A)Y(t) + B(v(t), v(t)) \qquad (5.6)$$

Here $Y'_+(\cdot)$ is the right derivative.

Now for $v \in C([0,T^*); D(A^\alpha))$ we have $B(v,v) \in C([0,T^*); X), AY \in C([0,T^*); D(A^{\alpha-1}))$ and $L_u Y \in C([0,T^*); D(A^{\alpha-1}))$ due to the isomorphism property of A and the estimates on the trilinear form $b(\cdot,\cdot,\cdot)$. Hence the right derivative $Y'_+(\cdot) \in C(0,T^*; D(A^{\alpha-1}))$. We thus conclude as in the chapter 4 that $Y'(\cdot) \in C(0,T^*; D(A^{\alpha-1}))$ and this proves the theorem.

♣

The existence and uniqueness of the solution of 5.2 can be obtained using the fixed point theoretic method used in [23]. If we seek the solution in a ball $B_\delta(0) \subset D(A^\alpha)$, the fixed point arguement will work for a maximal time interval $[0, T^*(v_0)[$, with $T^*(v_0) \le \infty$. The uniqueness result can in fact be realized in the following way: Suppose $v_1(t), v_2(t)$ be solutions corresponding to the initial data $v_0 \in D(A^\alpha)$. We then get

$$v_2(t) - v_1(t) = \int_0^t Z(t,\eta)[B(v_1,v_1) - B(v_2,v_2)]d\eta.$$

That is

$$\|v_2 - v_1\|_{C(0,t;D(A^\alpha))} \le y_1(t)\|v_2 - v_1\|_{C(0,t;D(A^\alpha))}\{\|v_1\|_{C(0,T^*;D(A^\alpha))} + \|v_2\|_{C(0,T^*;D(A^\alpha))}\}.$$

Here $y_1(t) \to 0$ as $t \to 0$ and hence the only possibility is that $v_1 = v_2$.

We will now establish an important result regarding the dependence of the solution on the initial data. Let us rewrite (5.2) in the form $\mathcal{F}(v,v_0) = 0$. Here the map

$$\mathcal{F}(\cdot,\cdot) : C(0,T^*; D(A^\alpha)) \times D(A^\alpha) \to C(0,T^*; D(A^\alpha))$$

is defined by

$$\mathcal{F}(v,v_0) = v - Z(t,0)v_0 + M(v,v). \qquad (5.7)$$

Here the bilinear operator $M(\cdot,\cdot) : C(0,T^*;D(A^\alpha))^{\otimes 2} \to C(0,T^*;D(A^\alpha))$ satisfies

$$M(0,w) = M(w,0) = 0, \forall w \in C(0,T^*;D(A^\alpha)) \text{ and}$$

$$Z(t,0) \in \mathcal{L}(D(A^\alpha),D(A^\alpha)) \text{ for fixed } t.$$

$\clubsuit$

Theorem 5.2 *For fixed $T_1 < T^*(v_0) \leq \infty$ there exists a neighborhood of the origin $B_\delta \subset D(A^\alpha)$ such that for $v_0 \in B_\delta$, there exists a unique solution $v \in C(0,T_1;D(A^\alpha))$ to the problem 5.2. Moreover, the dependence of v in the initial data v_0 is Frechet-analytic.*

Proof: Let us first establish the

Lemma 5.1 *The map $\mathcal{F}(\cdot,\cdot)$ is Frechet-analytic in a ball centered at the origin of*

$$C(0,T_1;D(A^\alpha)) \times D(A^\alpha)$$

Proof: Let us first compute the Frechet derivatives of $\mathcal{F}(\cdot,\cdot)$ with respect to v and v_0. Consider the balls $B_{1\rho}$ and $B_{2\rho}$ defined by

$$B_{1\rho} = \{v; v \in C(0,T_1;D(A^\alpha)); \|v\|_{C(0,T_1;D(A^\alpha))} \leq \rho\} \text{ and}$$

$$B_{2\rho} = \{v; v \in D(A^\alpha); \|v\|_{D(A^\alpha)} \leq \rho\}.$$

Now the G- derivatives of $\mathcal{F}(\cdot,\cdot)$ at $(\bar{v},\bar{v}_0) \in B_{1\rho} \times B_{2\rho}$ are given by:

$$[\delta_v^{(1)}\mathcal{F}](\bar{v},\bar{v}_0;h_1,0) = h_1 + M(\bar{v},h_1) + M(h_1,\bar{v})$$

$$[\delta_{v_0}^{(1)}\mathcal{F}](\bar{v},\bar{v}_0;0,h_2) = -Z(t,0)h_2$$

$$[\delta_v^{(2)}\mathcal{F}](\bar{v},\bar{v}_0;h_1,0) = 2M(h_1,h_1),$$

$$\forall h_1 \in C(0,T_1;D(A^\alpha)) \text{ and } \forall h_2 \in D(A^\alpha).$$

All the higher order derivatives are zero.

Let us note that setting $v = \bar{v} + h_1$ and $v_0 = \bar{v_0} + h_2$ in (5.7) gives

$$\mathcal{F}(\bar{v} + h_1, \bar{v_0} + h_2) = \{\bar{v} - Z(t,0)\bar{v_0} + M(\bar{v},\bar{v})\}$$

$$+h_1 - Z(t,0)h_2 + M(\bar{v},h_1) + M(h_1,\bar{v}) + M(h_1,h_1).$$

That is

$$\mathcal{F}(\bar{v} + h_1, \bar{v_0} + h_2) = \mathcal{F}(\bar{v},\bar{v_0}) + [\delta_v^{(1)}\mathcal{F}](\bar{v},\bar{v_0}; h_1, 0)$$

$$[\delta_{v_0}^{(1)}\mathcal{F}](\bar{v},\bar{v_0}; 0, h_2) + \frac{1}{2}[\delta_v^{(2)}\mathcal{F}](\bar{v},\bar{v_0}; h_1, 0). \tag{5.8}$$

Also note that

$$\|[\delta_v^{(1)}\mathcal{F}](\bar{v},\bar{v_0}; h_1, 0)\|_{C(0,T_1;D(A^\alpha))} \le (1 + \rho \text{ .const.})\|h_1\|_{C(0,T_1;D(A^\alpha))}$$

and this derivative is homogeneous of degree one in h_1. Thus

$$[\delta_v^{(1)}\mathcal{F}](\bar{v},\bar{v_0}; \cdot, 0) \in \mathcal{L}(C(0,T_1;D(A^\alpha)); C(0,T_1;D(A^\alpha))).$$

Similarly,

$$[\delta_{v_0}^{(1)}\mathcal{F}](\bar{v},\bar{v_0}; 0, \cdot) \in \mathcal{L}(D(A^\alpha), D(A^\alpha))$$

and is homogeneous of degree one in h_2. $[\delta_v^{(2)}\mathcal{F}](\bar{v},\bar{v_0}; h_1, 0)$ is homogeneous of degree two in h_1 and

$$\|[\delta_v^{(2)}\mathcal{F}](\bar{v},\bar{v_0}; h_1, 0)\|_{C(0,T_1;D(A^\alpha))} \le \text{ const.}\|h_1\|^2_{C(0,T_1;D(A^\alpha))}.$$

That is

$$[\delta_v^{(2)}\mathcal{F}](\bar{v},\bar{v_0}; \cdot, 0) : C(0,T_1;D(A^\alpha))^{\otimes 2} \to C(0,T_1;D(A^\alpha))$$

continuously. We conclude from these results that the map $\mathcal{F}(\cdot,\cdot)$ is Frechet analytic in a neighborhood of the origin of $C(0,T_1;D(A^\alpha)) \times D(A^\alpha)$.

♣

Let us now set $\bar{v} = \bar{v_0} = 0$. We then get, $[\delta_v^{(1)}\mathcal{F}](0,0; \cdot, 0) = 1$. Hence the Frechet derivative of the map $\mathcal{F}(\cdot,\cdot)$ with respect to the first variable v has continuous inverse at the origin. Moreover $\mathcal{F}(0,0) = 0$. Thus by the F-analytic version of the implicit function theorem[18] there exists a neighborhood of the origin $B_{1\delta} \times B_{2\delta} \subset B_{1\rho} \times B_{2\rho}$ and a map

$$W(t,0;\cdot) : B_{2\delta} \subset D(A^\alpha) \to B_{1\delta} \subset C(0,T_1;D(A^\alpha))$$

such that $v(t) = W(t,0;v_0)$ and $W(t,0;\cdot)$ is F-analytic in this neighborhood. That is $v(t)$ can be written as a power series in the initial data in the following way:

$$v(t) = W(t,0;v_0) = \sum_{n \geq 1} \mathcal{H}_n(v_0,\cdots,v_0;t). \tag{5.9}$$

The n-linear maps

$$\mathcal{H}_n(\cdots) : B_{2\delta}^{\otimes n} \to B_{1\delta} \text{ continuously} .$$

This series representation converges in the neighborhood defined above. One can verify easily that $\mathcal{H}_1(v_0;t) = Z(t,0)v_0$. We finally note that due to the uniqueness theorem,

$$W(t,0;v_0) = W(t,t_1;W(t_1,0;v_0)) \text{ for } 0 \leq t_1 \leq t < T^*(v_0)$$

and $W(0,0;\cdot) = I$.

If the basic flow field $U(t)$ is T-periodic then

$$W(2T,0;v_0) = W(T,0;W(T,0;v_0))$$

and in general

$$W(nT,0;\cdot) = W(T,0;\cdot)^n, 1 \leq n < \frac{T^*(v_0)}{T}. \tag{5.10}$$

As noted earlier the Frechet derivative of the map of the solution map $(DW)(t,0;\cdot) = Z(t,0)$ satisfies a similar relationship.

♣

Remark:

In three dimensions $T^*(v_0)$ can be taken to infinity by considering the data in a sufficiently small neighborhood. In two dimension this is possible without any restriction on data.

5.1 Nonlinear Semigroup For Unbounded Domains

One of the key steps in analyzing the nonlinear semigroup for unbounded domains is the careful characterization of the inertia terms $\boldsymbol{u} \cdot \nabla \boldsymbol{u}$. Let us now provide an analysis of these terms. We will observe that the bilinear map associated with the inertial term has the property

$$B(\cdot,\cdot) : D(A^{1/4}) \times D(A^{1/2}) \to D(A^{-1/4}) \text{ for } \Omega \subset \boldsymbol{R}^2$$

and

$$B(\cdot,\cdot) : D(A^{1/2}) \times D(A^{1/2}) \to D(A^{-1/4}) \text{ for } \Omega \subset \boldsymbol{R}^3.$$

The result for the two dimensional case was noted by Sobolevskii[78] and the three dimensional result is due to Masuda [60]. We will describe the proof for the two dimensional case below (since the other case is similar).

Let us begin with certain sharp estimate for the kernel of the fractional powers of the Friedrich's extension of the Laplacian $(-\Delta)$ defined for C^∞ vectorfields with compact support in Ω. The characterization of this operator is similar to that of the Stokes operator in unbounded domains (given in chapter 2.)

We begin with the bilinear form

$$a(\boldsymbol{u}, \boldsymbol{v}) = (\nabla \boldsymbol{u}, \nabla \boldsymbol{v})_{L^2(\Omega)}.$$

Now, for a given $\boldsymbol{u} \in H_0^1(\Omega)$, if there exists $\boldsymbol{g} \in L^2(\Omega)$ such that

$$a(\boldsymbol{u}, \boldsymbol{v}) = (g, \boldsymbol{v})_{L^2(\Omega)}, \forall \boldsymbol{u}, \boldsymbol{v} \in H_0^1(\Omega)$$

then we say that $\boldsymbol{u} \in D(\hat{\Delta})$ and set $g = \hat{\Delta}\boldsymbol{u} \in R(\hat{\Delta})$. We then get the continuous dense embedding

$$D(\hat{\Delta}) \subset H_0^1(\Omega) \subset L^2(\Omega) = L^2(\Omega)' \subset H^{-1}(\Omega) \subset D(\hat{\Delta})'$$

As before we conclude that $\hat{\Delta}$ is a closed operator. Moreover $D(\hat{\Delta}^{1/2}) = H_0^1(\Omega)$ with

$$\|\hat{\Delta}^{1/2}\boldsymbol{u}\|_{L^2(\Omega)} = \|\nabla\boldsymbol{u}\|_{L^2(\Omega)}, \forall \boldsymbol{u} \in D(\hat{\Delta}^{1/2}).$$

In order to characterize the fractional powers we will use as before $\hat{\Delta}_\epsilon = L + \epsilon I, 0 < \epsilon < 1$
We have in fact $\hat{\Delta}_\epsilon \in \mathcal{L}(D(\hat{\Delta}); L^2(\Omega)) \cap \mathcal{L}(H_0^1(\Omega); H^{-1}(\Omega))$ onto and

$$D(\hat{\Delta}_\epsilon^\alpha) = D(\hat{\Delta}^\alpha), \text{ for } \alpha \in (0,1).$$

Moreover

$$\|\hat{\Delta}_\epsilon^{1/2} u\|_{L^2(\Omega)}^2 = \|\nabla u\|_{L^2(\Omega)}^2 + \epsilon \|u\|_{L^2(\Omega)}^2, \forall u \in D(\hat{\Delta}^{1/2})$$

We will now prove that

Theorem 5.3 *Let $\Omega \subset \mathbf{R}^2$ be an arbitrary open set with class C^2 boundary $\partial\Omega$. Then we have for $0 < \alpha < 1$*

$$[L_\epsilon^{-\alpha} u](x) = \int_\Omega \mathcal{K}_\alpha(x,y;\epsilon) u(y) dy$$

with the kernel satisfing the estimate,

$$0 \leq \mathcal{K}_\alpha(x,y;\epsilon) \leq \frac{C(\alpha)}{|x-y|^{2-2\alpha}}, \forall x,y \in \Omega. \tag{5.11}$$

Proof:

We will provide a proof using an estimate for the heat kernel, which seems to hold for the kind of unbounded domains considered in this book. The resolvent of $-\hat{\Delta}_\epsilon$ satisfies,

$$\|R(\lambda; -\hat{\Delta}_\epsilon)\|_{\mathcal{L}(L^2(\Omega); L^2(\Omega))} \leq (\epsilon + \Re\lambda)^{-1} \text{ for } \Re\lambda \geq 0. \tag{5.12}$$

Here $R(\lambda; -\hat{\Delta}_\epsilon) = (\hat{\Delta}_\epsilon + \lambda)^{-1}$. Similarly considering

$$\Im((\hat{\Delta}_\epsilon + \lambda)u, u)_{L^2(\Omega)} = \Im(a_\epsilon(u,u) + \lambda(u,u)_{L^2(\Omega)})$$

we get

$$\|(\hat{\Delta}_\epsilon + \lambda)u\|_{L^2(\Omega)} \geq \Im\lambda \|u\|_{L^2(\Omega)} \text{ for } \Im\lambda \neq 0.$$

Since $\hat{\Delta}_\epsilon + \lambda$ is on to, we get,

$$\|R(\lambda; -\hat{\Delta}_\epsilon)\|_{\mathcal{L}(L^2(\Omega); L^2(\Omega))} \leq |\Im\lambda|^{-1} \text{ for } \Im\lambda \neq 0. \tag{5.13}$$

Combining (5.12) and (5.13) we get

$$\|R(\lambda; -\hat{\Delta}_\epsilon)\|_{\mathcal{L}(L^2(\Omega); L^2(\Omega))} \leq \frac{2^{1/2}}{|\lambda + \epsilon|} \text{ for } \Re\lambda \geq 0.$$

Hence by analytic continuation we get

$$\|R(\lambda; -\hat{\Delta}_\epsilon)\|_{\mathcal{L}(L^2(\Omega);L^2(\Omega))} \leq \frac{M}{|\lambda + \epsilon|} \text{ for } \lambda \in \Sigma. \tag{5.14}$$

where $\Sigma = \{\lambda; \pi/2 < |\arg \lambda| < \pi\}$.

We will now use this estimate to characterize the semigroup generated by $-\hat{\Delta}_\epsilon$. The operator $-\hat{\Delta}_\epsilon$ is closed with $D(\hat{\Delta}_\epsilon)$ dense in $L^2(\Omega)$ and its resolvent satsfies (5.14). Hence by Hille-Yoshida theorem $-\hat{\Delta}_\epsilon$ generates a C^0-semigroup $T_\epsilon(t)$ in $L^2(\Omega)$ with

$$\|T_\epsilon(t)\|_{\mathcal{L}(L^2(\Omega);L^2(\Omega))} \leq e^{-\epsilon t} \text{ for } t > 0.$$

Moreover using the estimate (5.14) we can show that $T_\epsilon(t)$ is analytic in an acute sector containing the positive real axis. This gives us another representation for the fractional powers of $\hat{\Delta}_\epsilon$ [44]:

$$\hat{\Delta}_\epsilon^{-\alpha} = \frac{1}{\Gamma(\alpha)} \int_0^\infty t^{\alpha-1} T_\epsilon(t) dt. \tag{5.15}$$

Now note that $T_\epsilon(t)u_0$ solves

$$u_t - \Delta u + \epsilon u = 0 \text{ in } \Omega$$

$$u|_{\partial\Omega} = 0 \tag{5.16}$$

$$u(x) \to 0 \text{ as } |x| \to \infty \text{ and}$$

$$u(x,0) = u_0(x) \text{ in } \Omega.$$

The solution can be expressed as

$$u(x,t) = \int_\Omega G_\epsilon(x,y;t)u_0(y)dy.$$

Thus

$$\hat{\Delta}_\epsilon^{-\alpha} u_0 = \frac{1}{\Gamma(\alpha)} \int_0^\infty t^{\alpha-1}\{\int_\Omega G_\epsilon(x,y;t)u_0(y)dy\}dt. \tag{5.17}$$

Hence

$$\mathcal{K}_\alpha(x,y) = \frac{1}{\Gamma(\alpha)} \int_0^\infty t^{\alpha-1} G_\epsilon(x,y;t)dt. \tag{5.18}$$

Thus in order to estimate $\mathcal{K}_\alpha(x,y)$, we simply use the estimate of the heat kernel:

$$G_\epsilon(x,y,\tau) \leq \frac{H(\tau)}{4\pi\tau} e^{-\frac{|x-y|^2}{4\tau}}.$$

Let us now estimate the kernel $\mathcal{K}_\alpha(x,y)$ as

$$\mathcal{K}_\alpha(x,y) \le \frac{1}{\Gamma(\alpha)} \int_0^\infty \frac{t^{\alpha-1}}{4\pi t} e^{-\frac{|x-y|^2}{4t}} \, dt.$$

This gives us

$$\mathcal{K}_\alpha(x,y) \le \frac{C(\alpha)}{|x-y|^{2-2\alpha}}, \forall x,y \in \Omega$$

with

$$C(\alpha) = \frac{\pi^{1-\alpha}}{4^\alpha \Gamma(\alpha)^2 \sin \pi\alpha} \text{ for } 0 < \alpha < 1.$$

♣

Let us define a bilinear form

$$F_\epsilon[\cdot,\cdot] : (\boldsymbol{V}_1(\Omega) \cap L^\infty(\Omega))^{\otimes 2} \to \boldsymbol{H}(\Omega)$$

as

$$F_\epsilon[\boldsymbol{u},\boldsymbol{v}] = A_\epsilon^{-1/4} P_H(\boldsymbol{u}\cdot\nabla)\boldsymbol{v} = A_\epsilon^{-1/4} B(\boldsymbol{u},\boldsymbol{v})$$

with

$$\boldsymbol{u},\boldsymbol{v} \in \boldsymbol{V}_1(\Omega) \cap L^\infty(\Omega) \text{ and } 0 < \epsilon < 1.$$

Here $P_H : L^2(\Omega) \to \boldsymbol{H}(\Omega)$ is the orthogonal projection operator and A_ϵ is the regularly accretive operator defined in chapter 2.

Lemma 5.2 $\forall \epsilon$ *such that* $0 < \epsilon < 1$

$$\|F_\epsilon[\boldsymbol{u},\boldsymbol{v}]\|_{H(\Omega)} \le \text{ const.} \|A^{1/4}\boldsymbol{u}\|_{H(\Omega)} \|A^{1/2}\boldsymbol{v}\|_{H(\Omega)} \ \forall \boldsymbol{u} \in D(A^{1/4}) \text{ and } \forall \boldsymbol{v} \in D(A^{1/2}).$$

Proof:

We have the orthogonal projection map

$$P_H \in \mathcal{L}(L^2(\Omega); \boldsymbol{H}(\Omega)) \cap \mathcal{L}(H_0^1(\Omega); \boldsymbol{V}_1(\Omega)).$$

Its adjoint

$$\mathcal{I} \in \mathcal{L}(\boldsymbol{H}(\Omega); L^2(\Omega)) \cap \mathcal{L}(\boldsymbol{V}_1(\Omega); H_0^1(\Omega))$$

is defined as

$$(v, \mathcal{I}u)_{L^2(\Omega)} = (P_H v, u)_{L^2(\Omega)}, \quad \forall u \in \boldsymbol{H}(\Omega), \forall v \in L^2(\Omega).$$

$\mathcal{I}$ is the identity map of the embeddings $\boldsymbol{H}(\Omega) \subset L^2(\Omega)$ and $\boldsymbol{V}_1(\Omega) \subset H_0^1(\Omega)$. That is $\mathcal{I}(\boldsymbol{H}(\Omega)) \subset L^2(\Omega)$ and $\mathcal{I}(D(A^{1/2})) \subset D(\hat{\Delta}^{1/2})$. We also have, for $0 \le \epsilon < 1$

$$\|A_\epsilon^{1/2} u\|_{L^2(\Omega)} = \|\hat{\Delta}_\epsilon^{1/2} \mathcal{I}u\|_{L^2(\Omega)} \quad \forall u \in \boldsymbol{V}_1(\Omega). \tag{5.19}$$

We then deduce using the Heinz-Kato theorem[85, 44], for $0 < \alpha < 1/2$

$$\|\hat{\Delta}_\epsilon^\alpha \mathcal{I}u\|_{L^2(\Omega)} \le \|A_\epsilon^\alpha u\|_{L^2(\Omega)}, \quad \forall u \in D(A^\alpha) \tag{5.20}$$

and

$$\mathcal{I}(D(A^\alpha)) \subset D(\hat{\Delta}^\alpha).$$

Now note that, setting in (5.20) $u = A_\epsilon^{-\alpha} v$ with $v \in \boldsymbol{H}(\Omega)$ we get

$$\|\hat{\Delta}_\epsilon^\alpha \mathcal{I} A_\epsilon^{-\alpha} v\|_{L^2(\Omega)} \le \|v\|_{H(\Omega)}, \forall v \in \boldsymbol{H}(\Omega), 0 < \alpha < 1/2.$$

Thus the operator $S_\epsilon = \hat{\Delta}_\epsilon^\alpha \mathcal{I} A_\epsilon^{-\alpha}$ satisfies

$$\|S_\epsilon\|_{\mathcal{L}(H(\Omega);L^2(\Omega)} \le 1.$$

The adjoint of this operator $S_\epsilon^* = A_\epsilon^{-\alpha} P_H \hat{\Delta}_\epsilon^\alpha$ satisfies

$$\|S_\epsilon\|_{\mathcal{L}(H(\Omega);L^2(\Omega))} = \|S_\epsilon^*\|_{\mathcal{L}(L^2(\Omega);H(\Omega))} \le 1. \tag{5.21}$$

Here the adjoint is defined as

$$(S_\epsilon u, v)_{L^2(\Omega)} = (u, S_\epsilon^* v)_{L^2(\Omega)}, \forall u \in D(S_\epsilon) = \boldsymbol{H}(\Omega) \text{ and } \forall v \in D(S_\epsilon^*) = L^2(\Omega).$$

Note here that S_ϵ^* is actually an extention of the operator $A_\epsilon^{-\alpha} P_H \hat{\Delta}_\epsilon^\alpha$ which is defined in $D(\hat{\Delta}_\epsilon^\alpha)$.

Let us now set $g(x) = (\boldsymbol{u} \cdot \nabla)\boldsymbol{v}$ for $\boldsymbol{u}, \boldsymbol{v} \in \boldsymbol{V}_1(\Omega) \cap L^\infty(\Omega)$ and consider

$$\|\hat{\Delta}_\epsilon^{-1/4} g\|_{L^2(\Omega)}^2 = (\hat{\Delta}_\epsilon^{-1/4} g, \hat{\Delta}_\epsilon^{-1/4} g)_{L^2(\Omega)}.$$

Since $\hat{\Delta}_\epsilon^{-1/4}$ is self adjoint we get

$$\|\hat{\Delta}_\epsilon^{-1/4}g\|_{L^2(\Omega)}^2 = (\hat{\Delta}_\epsilon^{-1/2}g,g)_{L^2(\Omega)} = \sum_{i,j=1}^{2}\int_\Omega\{\int_\Omega \mathcal{K}_{1/2}^{ij}(x,y)g^j(y)dy\}g^i(x)dx.$$

From the estimate obtained earlier (for the scalar case) for the kernel we have

$$\mathcal{K}_{1/2}^{ij}(x,y) \leq \frac{C(1/2)}{|x-y|} \text{ for } i,j=1,2$$

Now

$$|\sum_{ij=1}^{2}\mathcal{K}_{1/2}^{ij}(x,y)g^i(y)g^j(x)| \leq \frac{C(1/2)}{|x-y|}\sum_{i,k=1}^{2}|\frac{\partial v^i}{\partial y^k}||u^k(y)|\sum_{j,l=1}^{2}|\frac{\partial v^j}{\partial x^l}||u^l(x)|$$

$$\leq \frac{C(1/2)}{|x-y|}\{(\sum_{k=1}^{2}u_k(y)^2)\sum_{i,k=1}^{2}|\frac{\partial v^i}{\partial y^k}|^2\}^{1/2}\{(\sum_{l=1}^{2}u_l(x)^2)\sum_{j,l=1}^{2}|\frac{\partial v^j}{\partial x^l}|^2\}^{1/2}.$$

Thus

$$\|\hat{\Delta}_\epsilon^{-1/4}g\|_{L^2(\Omega)}^2 \leq C(1/2)\int_\Omega\int_\Omega\frac{1}{|x-y|}|u(x)||u(y)||\nabla v(y)||\nabla v(x)|dxdy.$$

Applying Schwartz inequality we get

$$\leq C(1/2)\{\int_\Omega\int_\Omega\frac{1}{|x-y|}|u(x)|^2|\nabla v(y)|^2dxdy\}^{1/2}\{\int_\Omega\int_\Omega\frac{1}{|x-y|}|u(y)|^2|\nabla v(x)|^2dxdy\}^{1/2}.$$

We will show later that $\forall y \in \Omega$ and $\forall u \in D(\hat{\Delta}^{1/4})$,

$$\int_\Omega\frac{1}{|x-y|}|u(x)|^2dx \leq \text{ const.}\|\hat{\Delta}^{1/4}u\|_{L^2(\Omega)}. \tag{5.22}$$

Thus noting that $\|\hat{\Delta}^{1/2}u\|_{L^2(\Omega)} = \|\nabla u\|_{L^2(\Omega)}$ we get

$$\|\hat{\Delta}_\epsilon^{-1/4}g\|_{L^2(\Omega)} \leq \text{ const. }\|\hat{\Delta}^{1/2}v\|_{L^2(\Omega)}\|\hat{\Delta}^{1/4}u\|_{L^2(\Omega)}. \tag{5.23}$$

We will now show (5.22). Let us consider $u \in D(\hat{\Delta}^{1/4})$ and set $w_\epsilon = \hat{\Delta}_\epsilon^{1/4}u$ for some $\epsilon, 0 < \epsilon < 1$. Then $u = \hat{\Delta}_\epsilon^{-1/4}w$ and using the estimate (5.11) for the kernel we get

$$|u(x)| \leq \text{ const. }\int_\Omega\frac{1}{|x-y|^{3/2}}|w(y)|dy.$$

Let $z \in \Omega$ be an arbitrary point then by Schwartz inequality,

$$|u(x)| \leq \text{ const. }\{\int_\Omega\frac{|x-z|^{1/2}}{|y-z||x-y|^{3/2}}dy\}^{1/2}\{\int_\Omega\frac{|y-z||w(y)|^2}{|x-z|^{1/2}|x-y|^{3/2}}dy\}^{1/2}.$$

Noting that the first integral on the right is bounded independent of $z \in \Omega$ (this is easy to show for the case of exterior domains if we switch to polar coordinate system) we get

$$\int_\Omega \frac{1}{|x-z|}|u(x)|^2 dx \leq \text{const.}\{\int_\Omega |w(y)|^2 dy\}\{\int_\Omega \frac{|y-z|}{|x-z|^{3/2}|x-y|^{3/2}}dx\}.$$

The second integral is again bounded independent of $y \in \Omega$. Thus,

$$\int_\Omega \frac{1}{|x-z|}|u(x)|^2 dx \leq \text{const.}\{\int_\Omega |w(y)|^2 dy\} = \text{const.} \ \|\hat{\Delta}_\epsilon^{1/4} u\|_{L^2(\Omega)}.$$

This gives us

$$\int_\Omega \frac{1}{|x-z|}|u(x)|^2 dx \leq \text{const.}\|\hat{\Delta}_\epsilon^{1/4} u\|_{L^2(\Omega)} \text{ for } 0 < \epsilon < 1.$$

Noting that ϵ is arbitrary and that $\hat{\Delta}_\epsilon^{1/4} \to \hat{\Delta}^{1/4}$ strongly we get (5.22).

Let us now proceed with the proof of the lemma. We will write

$$F_\epsilon[u,v] = A_\epsilon^{-1/4} P_H \hat{\Delta}_\epsilon^{1/4} \hat{\Delta}_\epsilon^{-1/4}(u \cdot \nabla)v = S_\epsilon^* \hat{\Delta}_\epsilon^{-1/4} g.$$

Thus using (5.23) and (5.21),

$$\|F_\epsilon[u,v]\|_{L^2(\Omega)} \leq \|S_\epsilon^*\|_{\mathcal{L}(H(\Omega);L^2(\Omega))}\|\hat{\Delta}_\epsilon^{-1/4} g\|_{L^2(\Omega)}$$

$$\leq 1. \text{ const. } \|\hat{\Delta}^{1/2}v\|_{L^2(\Omega)}\|\hat{\Delta}^{1/4}u\|_{L^2(\Omega)}.$$

Now using (5.20) we get

$$\|F_\epsilon[u,v]\|_{L^2(\Omega)} \leq \text{const.} \ \|A^{1/2}v\|_{L^2(\Omega)}\|A^{1/4}u\|_{L^2(\Omega)}.$$

♣

We will now prove

Theorem 5.4 *(i) For each fixed $u,v \in V_1(\Omega) \cap L^\infty(\Omega)$, $F_\epsilon[u,v]$ converges to some element $\tilde{F}[u,v]$ in the weak topology of $H(\Omega)$ as $\epsilon \to 0$.*

(ii) $\tilde{F}[u,v]$ thus defined can be extended to the bilinear operator $F[u,v] \in \mathcal{L}(D(A^{1/4}) \times D(A^{1/2}); H(\Omega))$ with the estimates

$$(a)\|F[u,v]\|_{L^2(\Omega)} \leq M\|A^{1/4}u\|_{L^2(\Omega)}\|A^{1/2}v\|_{L^2(\Omega)} \quad \forall v \in D(A^{1/2}) \text{ and } \forall u \in D(A^{1/4})$$

and

$$(b)\|F[u_1, v_1] - F[u_2, v_2]\|_{L^2(\Omega)} \le M[\|A^{1/4}u_1 - A^{1/4}u_2\|_{L^2(\Omega)}\|A^{1/2}v_1\|_{L^2(\Omega)}$$

$$+\|A^{1/4}u_2\|_{L^2(\Omega)}\|A^{1/2}v_1 - A^{1/2}v_2\|_{L^2(\Omega)}] \ \forall v_1, v_2 \in D(A^{1/2}) \ and \ \forall u_1, u_2 \in D(A^{1/4}).$$

Proof:

Let $u, v \in V_1(\Omega) \cap L^\infty(\Omega)$. Then $P_H(u \cdot \nabla)v = B(u, v) \in H(\Omega)$. Let us take $\phi \in D(A^{1/2})$ and consider

$$(F_\epsilon[u, v], A^{1/4}\phi)_{L^2(\Omega)} = (B(u, v), A_\epsilon^{-1/4}A^{1/4}\phi)_{L^2(\Omega)}.$$

Since $B(u, v) \in H(\Omega)$, we have

$$\lim_{\epsilon \to 0}(F_\epsilon[u, v], A^{1/4}\phi)_{L^2(\Omega)} = (B(u, v), \phi)_{L^2(\Omega)}. \tag{5.24}$$

Here we used the fact that $\forall \phi \in D(A^{1/4})$ and $\forall \psi \in H(\Omega)$,

$$\lim_{\epsilon \to 0}(\psi, A_\epsilon^{-1/4}A^{1/4}\phi)_{L^2(\Omega)} = (\psi, \phi)_{L^2(\Omega)}.$$

In fact

$$|(\psi, (A_\epsilon^{-1/4}A^{1/4} - I)\phi)_{L^2(\Omega)}| \le \|\psi\|_{L^2(\Omega)}\|(A_\epsilon^{-1/4}A^{1/4} - I)\phi\|_{L^2(\Omega)}.$$

Now,

$$\|\phi - A_\epsilon^{-1/4}A^{1/4}\phi\|_{L^2(\Omega)} = \|A_\epsilon^{-1/4}(A_\epsilon^{1/4} - A^{1/4})\phi\|_{L^2(\Omega)}.$$

Noting that $A_\epsilon^{-1/4}$ is a bounded operator we get using theorem 2.11 that,

$$|(\psi, (A_\epsilon^{-1/4}A^{1/4} - I)\phi)_{L^2(\Omega)}| \le \ \text{const.} \ \epsilon^{1/4}\|\phi\|_{L^2(\Omega)}\|\psi\|_{L^2(\Omega)}, \forall \phi \in D(A^{1/4}), \forall \psi \in H(\Omega)$$

Taking $\epsilon \to 0$ we get the desired result.

We have thus proved that $F_\epsilon[u, v]$ converges to some element $\tilde{F}[u, v]$ in the following sense:

$$\lim_{\epsilon \to 0}(F_\epsilon[u, v], \psi)_{L^2(\Omega)} = (\tilde{F}[u, v], \psi)_{L^2(\Omega)}, \ \ \forall \psi \in R(A^{1/4}).$$

Note that $F_\epsilon[u, v]$ is uniformly bounded for $0 < \epsilon < 1$ by lemma 5.2, and that $R(A^{1/4}) \subset H(\Omega)$ is dense. From this we can deduce the weak convergence of in $H(\Omega)$ in the following way: let $\psi \in H(\Omega)$. then (due to the density of $R(A^{1/4})$) for a given $\delta > 0, \exists \hat{\psi}$ such that

$\|\psi - \hat{\psi}\|_{L^2(\Omega)} < \delta.$

Now

$$|(F_\epsilon[u,v] - \tilde{F}[u,v], \psi)_{L^2(\Omega)}| \leq |(F_\epsilon[u,v] - \tilde{F}[u,v], \hat{\psi})_{L^2(\Omega)}| + |(F_\epsilon[u,v] - \tilde{F}[u,v], \psi - \hat{\psi})_{L^2(\Omega)}|$$

$$\leq C_1 \epsilon^{1/4} + C_2 \delta.$$

Taking $\epsilon \to 0$ and noting that $\delta > 0$ is arbitrary, we get the weak convergence in $\boldsymbol{H}(\Omega)$. Note that $\forall \psi \in \boldsymbol{H}(\Omega)$,

$$|(\tilde{F}[u,v], \psi)_{L^2(\Omega)}| \leq |(\tilde{F}[u,v] - F_\epsilon[u,v], \psi)_{L^2(\Omega)}| + |(F_\epsilon[u,v], \psi)_{L^2(\Omega)}|.$$

Set in particular $\psi = \tilde{F}[u,v]$ and let $\epsilon \to 0$, to get,

$$\|\tilde{F}[u,v]\|_{L^2(\Omega)} \leq M\|A^{1/4}u\|_{L^2(\Omega)}\|A^{1/2}v\|_{L^2(\Omega)}, \forall u \in D(A^{1/4}) \text{ and } \forall v \in D(A^{1/2}).$$

The operator $\tilde{F}[u,v]$ is bilinear since it is the limit of bilinear operators. Hence

$$\|\tilde{F}[u_1,v_1] - \tilde{F}[u_2,v_2]\|_{L^2(\Omega)} \leq \|\tilde{F}[u_1-u_2,v_1]\|_{L^2(\Omega)} + \|\tilde{F}[u_2,v_1-v_2]\|_{L^2(\Omega)}$$

$$\leq [\|A^{1/4}u_1 - A^{1/4}u_2\|_{L^2(\Omega)}\|A^{1/2}v_1\|_{L^2(\Omega)}$$

$$+ \|A^{1/4}u_2\|_{L^2(\Omega)}\|A^{1/2}v_1 - A^{1/2}v_2\|_{L^2(\Omega)} \quad \forall v_1, v_2 \in D(A^{1/2}) \text{ and } \forall u_1, u_2 \in D(A^{1/4}).$$

$\tilde{F}[u,v]$ thus characterized is defined in $(\boldsymbol{V}_1(\Omega) \cap L^\infty(\Omega))^{\otimes 2}$. Let us now extend this operator to $D(A^{1/4}) \times D(A^{1/2})$.

Let $u \in D(A^{1/4})$ and $v \in D(A^{1/2})$. Then due to the density of $D(A^{1/2}) \cap L^\infty(\Omega) \subset D(A^{1/4})$ and $D(A^{1/2}) \cap L^\infty(\Omega) \subset D(A^{1/2})$ there exists sequences $\{u_n\}, \{v_n\} \in D(A^{1/2})$ converging strongly to u and v respectively. Thus we can characterize the strong limit of $\tilde{F}[u,v]$ as

$$F[u,v] = \text{strong } \lim_{n \to \infty} \tilde{F}[u_n, v_n].$$

The bilinear operator $F[\cdot,\cdot] : \mathcal{L}(D(A^{1/4}) \times D(A^{1/2}); \boldsymbol{H}(\Omega))$ thus defined will obviously satisfy the estimates (a) and (b) due to the corresponding estimates for $\tilde{F}[u_n, v_n]$ and due to the strong convergence of u_n to u in the norm of $D(A^{1/4})$ and v_n to v in the norm of $D(A^{1/2})$.

$$\clubsuit$$

Chapter 6

Invariant Cone Theorem

We will show in this chapter that under certain hypothesis on the spectrum of $Z(T,0)$, there is a double cone at the origin of $D(A^\alpha), \alpha \in [1/2,1]$ that is locally invariant to the action of the map $W(T,0;\cdot)$. This result will enable us to study the characteristics of the orbits nearby a given periodic orbit. We will establish in particular the stability and instability properties of the original orbit. For similar stability theorems in specific topologies see[95, 73, 40]. *The central theorem proved below also contains the Principle of linearized stability for hydrodynamic transition.* The analysis of this chapter has been influenced by the results on general mappings in Banach spaces by Kirchgassner and Scheurle[75, 43].

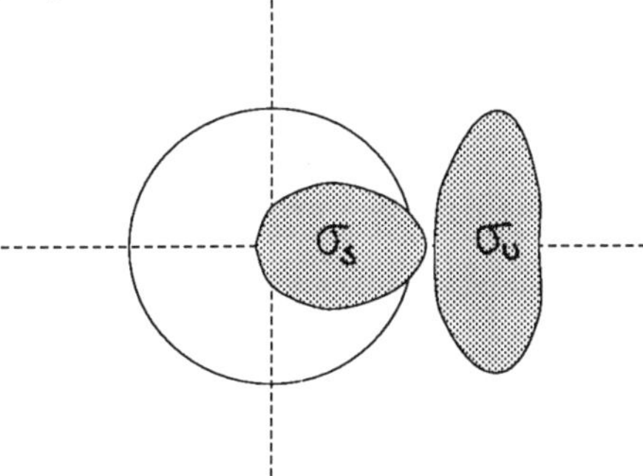

Figure 6.1 Spectrum of $Z(T,0)$

Theorem 6.1 *Let the basic flow be T-periodic in time so that the solution map satisfies (5.10).*

(i) If the spectrum of the monodromy operator $Z(T,0)$ lies inside the unit disc (spectral radius < 1) then the basic periodic solution is (locally) exponentially stable: there exists $\rho > 0$ such that $\forall v_0 \in B_\rho(0), W(t,0;v_0) \to 0$ exponentially in the norm of $D(A^\alpha), \alpha \in [1/2,1]$.

(ii) Let the spectrum of $Z(T,0)$ splits in to two disjoint sets σ_u and σ_s such that (figure 6.1),

$$\sup_{\sigma_s} |\lambda| < \inf_{\sigma_u} |\lambda| = a, a > 1 \ holds \ .$$

Then there exists a double cone $\mathcal{K} \subset D(A^\alpha)$ and a ball $B_\delta(0) \subset D(A^\alpha)$ such that $\forall v_0 \in B_\delta(0) \cap \mathcal{K}\backslash\{0\}$, there exists $n \in \mathcal{N}$ for which $\|W(nT,0;v_0)\|_{D(A^\alpha)} > \delta$. That is the basic solution is Lyapunov unstable.

Proof

Let us first state the following fundamental result:

Proposition 6.1 *Let the spectrum of the monodromy operator $Z(T,0) \in \mathcal{L}(D(A^\alpha); D(A^\alpha))$ splits in to two disjoint sets σ_u and σ_s such that $\sigma(Z(T,0)) = \sigma_u \cup \sigma_s$ and*

$$b = \sup_{\sigma_s} |\lambda| < \inf_{\sigma_u} |\lambda| = a.$$

Let P_u and P_s be the spectral projectors defined by the Dunford's integrals,

$$P_u = \frac{1}{2\pi i} \int_{\Gamma_u} R(\lambda; Z(T,0)) d\lambda \ and$$

$$P_s = \frac{1}{2\pi i} \int_{\Gamma_s} R(\lambda; Z(T,0)) d\lambda.$$

(Here $R(\lambda; Z(T,0))$ is the resolvent operator and Γ_u, Γ_s encircle σ_u, σ_s respectively). Then $\forall \epsilon > 0$, we can choose a norm in $D(A^\alpha)$, equivalent to the given one such that, $\forall v \in D(A^\alpha)$

$$(i) \ \|v\|_{D(A^\alpha)} = \|P_u v\|_{D(A^\alpha)} + \|P_s v\|_{D(A^\alpha)}, \ \ \|P_u\|_{D(A^\alpha)} = \|P_s\|_{D(A^\alpha)} = 1$$

$$(ii) \ \|Z(T,0)P_s v\|_{D(A^\alpha)} \leq (b + \epsilon)\|P_s v\|_{D(A^\alpha)}$$

$$(iii) \ \|Z(T,0)P_u v\|_{D(A^\alpha)} \geq (a - \epsilon)\|P_u v\|_{D(A^\alpha)}.$$

Here $P_s + P_u = I$, $P_s P_u = P_u P_s$ and P_s, P_u commute with $Z(T,0)$.

This result is in fact valid for any continuous linear operator in a Banach space with the above spectral properties. For the proof of this results see[18].

Let us now prove the part (i) of the theorem. Note that in this case σ_u is empty and σ_s lies inside the unit disc. Hence due to the proposition 6.1 we can renorm $D(A^\alpha)$ such that,

$$\|Z(T,0)\|_{\mathcal{L}(D(A^\alpha),D(A^\alpha))} \leq k < 1.$$

Now, the above result and the lemma 4.14 imply that,

$$\|Z(T+t_0,t_0)\|_{\mathcal{L}(D(A^\alpha),D(A^\alpha))} \leq k < 1, \forall t_0 \geq 0.$$

Note that the F-analyticity of the map $W(T+t_0,t_0;\cdot)$ implies that $\forall \epsilon > 0$, there exists $\delta > 0$ such that ,

$$\|W(T+t_0,t_0;v_0) - Z(T+t_0,t_0)v_0\|_{D(A^\alpha)} \leq \epsilon\|v_0\|_{D(A^\alpha)} \text{ for } \|v_0\|_{D(A^\alpha)} < \delta.$$

Hence

$$\|W(T+t_0,t_0;v_0)\|_{D(A^\alpha)} \leq (k+\epsilon)\|v_0\|_{D(A^\alpha)}.$$

Now the equation (5.10) gives

$$v(nT+t_0) = W(nT+t_0,t_0;v(t_0)) = W(T+t_0,t_0;\cdots,\cdots,W(T+t_0,t_0;v(t_0))$$

We thus obtain, $\forall n \geq 1$,

$$\|W(nT+t_0,t_0;v(t_0))\|_{D(A^\alpha)} \leq (k+\epsilon)^n\|v(t_0)\|_{D(A^\alpha)}$$

$$\leq \exp\{\frac{nT}{T}\ln(k+\epsilon)\}\|v(t_0)\|_{D(A^\alpha)}. \tag{6.1}$$

Let us now consider arbitrary t and choose n such that $0 \leq t - nT = t_0 < T$. We have,

$$v(t_0) = W(t_0,0;v_0) = Z(t_0,0)v_0 - \int_0^{t_0} Z(t_0,\eta)B(v(\eta),v(\eta))d\eta.$$

Using the estimates for

$$Z(t_0,\eta) \in \mathcal{L}(D(A^\alpha);D(A^\alpha)) \cap \mathcal{L}(E_1;D(A)) \cap \mathcal{L}(D(A^{-1/4});D(A^{1/2}))$$

obtained in the chapter 4 we get,

$$\|v(t_0)\|_{D(A^\alpha)} \leq C_1(t_0, \|v_0\|_{D(A^\alpha)})\|v_0\|_{D(A^\alpha)} \tag{6.2}$$

where $C_1(t_0, \|v_0\|_{D(A^\alpha)})$ depends on t_0 and the norm of v_0 and is bounded. Combining estimates (6.1) and (6.2) we get $\forall t > 0$,

$$\|v(t)\|_{D(A^\alpha)} = \|W(nT + t_0, t_0; v(t_0))\|_{D(A^\alpha)} \leq C_2(t_0, \|v_0\|_{D(A^\alpha)})e^{-\sigma t}\|v_0\|_{D(A^\alpha)} \tag{6.3}$$

with $\sigma = -\frac{1}{T}\ln(k + \epsilon)$. We thus conclude that $W(t, 0; v_0) \to 0$ exponentially when $v_0 \in B_\delta(0) \subset D(A^\alpha)$.

Let us now consider the part (ii) of the theorem. Again utilizing the proposition 6.1 we get $\forall v \in D(A^\alpha)$,

$$\|v\|_{D(A^\alpha)} = \|P_u v\|_{D(A^\alpha)} + \|P_s v\|_{D(A^\alpha)}, \quad \|P_u\|_{D(A^\alpha)} = \|P_s\|_{D(A^\alpha)} = 1$$

$$\|Z(T, 0)P_u v\|_{D(A^\alpha)} \geq a\|P_u v\|_{D(A^\alpha)} \text{ and}$$

$$\|Z(T, 0)P_s v\|_{D(A^\alpha)} \leq (a - \rho)\|P_s v\|_{D(A^\alpha)}.$$

Here

$$\sup_{\lambda \in \sigma_s} |\lambda| < a - \rho < a < \inf_{\lambda \in \sigma_u} |\lambda| \text{ and } a > 1, \rho > 0.$$

Let us define a double cone by (see figure 6.2)

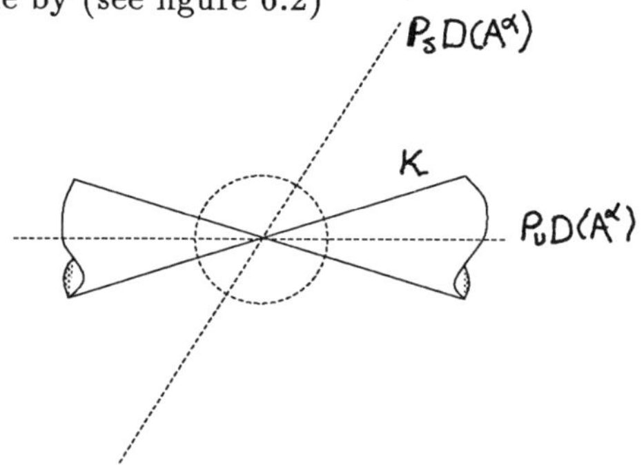

Figure 6.2 Double Cone $\mathcal{K}$

$$\mathcal{K} = \{v \in D(A^\alpha) \text{ such that } \|P_s v\|_{D(A^\alpha)} \leq q\|P_u v\|_{D(A^\alpha)}, q > 0\}, \tag{6.4}$$

$\dot{\mathcal{K}}= \mathcal{K}\backslash\{0\}$ and $\overset{\circ}{\mathcal{K}}=$ interior of $\mathcal{K}$. We will first establish the invariance properties of this cone with respect to $W(T,0;\cdot)$ and its Frechet derivatives.

Lemma 6.1 *Let the hypothesis of the part (ii) of the theorem holds. Then there exists a double cone $\mathcal{K}$ and a ball $B_\delta(0) \subset D(A^\alpha)$ such that*

(i) $(DW)(T,0;v_0) :\dot{\mathcal{K}}\to\overset{\circ}{\mathcal{K}}, \forall v_0 \in B_\delta(0)$ and the restriction of this Frechet derivative to $\dot{\mathcal{K}}$ is injective.

(ii) $W(T,0;\cdot) : B_\delta(0)\cap \dot{\mathcal{K}}\to\dot{\mathcal{K}}$ and the restriction of this map to $B_\delta(0)\cap \dot{\mathcal{K}}$ is injective.

(iii) $B_\delta(0)\cap \dot{\mathcal{K}} \cap M_s = empty, where$

$$M_s = \{v_0 \in B_\delta(0) \ such\ that\ , W(nT,0;v_0) \in B_\delta(0), \forall n\}$$

Proof:

Suppose $v_0 \in\dot{\mathcal{K}}$ then

$$\|P_s Z(T,0)v_0\|_{D(A^\alpha)} \le (a-\rho)\|P_s v_0\|_{D(A^\alpha)}$$

$$\le (a-\rho)q\|P_u v_0\|_{D(A^\alpha)}$$

$$\le (1-\frac{\rho}{a})q\|P_u Z(T,0)v_0\|_{D(A^\alpha)}.$$

Thus

$$\|P_s Z(T,0)v_0\|_{D(A^\alpha)} < q\|P_u Z(T,0)v_0\|_{D(A^\alpha)} \tag{6.5}$$

which implies that, $Z(T,0) :\dot{\mathcal{K}}\to\overset{\circ}{\mathcal{K}}$. Note also that for $v_0 \in\dot{\mathcal{K}}$

$$\|Z(T,0)v_0\|_{D(A^\alpha)} = \|P_u Z(T,0)v_0\|_{D(A^\alpha)} + \|P_s Z(T,0)v_0\|_{D(A^\alpha)}$$

$$\|Z(T,0)v_0\|_{D(A^\alpha)} \ge \|P_u Z(T,0)v_0\|_{D(A^\alpha)}$$

$$\ge a\|P_u v_0\|_{D(A^\alpha)} \ge \frac{a}{(1+q)}\|v_0\|_{D(A^\alpha)}.$$

Choosing small enough q we get,

$$\|Z(T,0)v_0\|_{D(A^\alpha)} \ge (1+\eta)\|v_0\|_{D(A^\alpha)}, \forall v_0 \in\dot{\mathcal{K}} \tag{6.6}$$

for some $\eta > 0$. This shows that $Z(T,0)|_{\dot{\mathcal{K}}}$ is injective.

Now, let $v_0 \in B_\delta(0)$ and denote the Frechet derivative of $W(T,0;\cdot)$ at v_0 as $(DW)(T,0;v_0)$. Then due to the analyticity of the map $W(T,0;\cdot)$ we have, $\forall \epsilon > 0$ there exists $\delta > 0$ such that

$$\|DW(T,0;v_0) - Z(T,0)\|_{\mathcal{L}(D(A^\alpha),D(A^\alpha))} < \epsilon, \forall v_0 \in B_\delta(0).$$

Hence, for $h \in \mathcal{K}$ and $v_0 \in B_\delta(0) \subset D(A^\alpha)$

$$\|DW(T,0;v_0)h - Z(T,0)h\|_{D(A^\alpha)} \leq \epsilon\|h\|_{D(A^\alpha)}$$

$$\leq \epsilon(1+q)\|P_u Z(T,0)h\|_{D(A^\alpha)}. \tag{6.7}$$

This implies that,

$$\|P_s DW(T,0;v_0)h - P_s Z(T,0)h\|_{D(A^\alpha)} \leq \epsilon(1+q)\|P_u Z(T,0)h\|_{D(A^\alpha)} \tag{6.8}$$

and

$$\|P_u DW(T,0;v_0)h - P_u Z(T,0)h\|_{D(A^\alpha)} \leq \epsilon(1+q)\|P_u Z(T,0)h\|_{D(A^\alpha)}. \tag{6.9}$$

Now let $\beta_1, \beta_2, \cdots, \beta_N \in \mathbf{R}$ be such that $\sum_{j=1}^{N} \beta_j = 1$. Let us define $L \in \mathcal{L}(D(A^\alpha), D(A^\alpha))$ as

$$L = \sum_{j=1}^{N} \beta_j (DW)(T,0;v_j) \text{ where } \{v_j\}_{j=1}^{N} \in B_\delta(0)$$

be arbitrary elements. Let us now estimate

$$\|P_u Lh\|_{D(A^\alpha)} \geq \|P_u Z(T,0)h\|_{D(A^\alpha)} - \left\|P_u \left\{\sum_{j=1}^{N} \beta_j (DW)(T,0;v_j) - Z(T,0)\right\} h\right\|_{D(A^\alpha)}.$$

Using equation (6.9) we get,

$$\|P_u Lh\|_{D(A^\alpha)} \geq (1 - \epsilon(1+q))\|P_u Z(T,0)h\|_{D(A^\alpha)}. \tag{6.10}$$

Similarly

$$\|P_s Lh\|_{D(A^\alpha)} \leq \|P_s Z(T,0)h\|_{D(A^\alpha)} + \left\|P_s \left\{\sum_{j=1}^{N} \beta_j (DW)(T,0;v_j) - Z(T,0)\right\} h\right\|_{D(A^\alpha)}.$$

Now using equation (6.8) we get

$$\|P_s Lh\|_{D(A^\alpha)} \leq (q - \sigma + \epsilon(1+q))\|P_u Z(T,0)h\|_{D(A^\alpha)} \text{ for } \sigma = \frac{\rho q}{a}. \tag{6.11}$$

Since $Z(T,0): \dot{\mathcal{K}} \to \overset{\circ}{\mathcal{K}}$ we have $\|P_u Z(T,0)h\|_{D(A^\alpha)} \neq 0$ for $h \in \dot{\mathcal{K}}$. Hence choosing $\epsilon > 0$ such that, $1 + \eta - \epsilon > 1, 1 - \epsilon(1+q) > 0$ we get

$$\|P_s Lh\|_{D(A^\alpha)} < q\|P_u Lh\|_{D(A^\alpha)}.$$

This implies that $L : \dot{\mathcal{K}} \to \overset{\circ}{\mathcal{K}}$. Moreover

$$\|Lh\|_{D(A^\alpha)} \geq \|Z(T,0)h\|_{D(A^\alpha)} - \left\| \left\{ \sum_{j=1}^{N} \beta_j (DW)(T,0;\boldsymbol{v}_j) - Z(T,0) \right\} h \right\|_{D(A^\alpha)}$$

Using equation (6.6)

$$\|Lh\|_{D(A^\alpha)} \geq (1 + \eta - \epsilon)\|h\|_{D(A^\alpha)} \quad \forall h \in \dot{\mathcal{K}} \tag{6.12}$$

Hence $\dot{\mathcal{K}}$ is invariant to the action of the map L and $L|_{\dot{\mathcal{K}}}$ is injective. Choosing $\beta_j = 0$ for all j except $j = 1$ and $\beta_1 = 1$ gives part (i) of the lemma.

Let $\boldsymbol{v}_0, \boldsymbol{u}_0 \in B_\delta(0)$ be such that $\boldsymbol{u}_0 - \boldsymbol{v}_0 \in \dot{\mathcal{K}}$. Then for $\lambda \in [0,1]$ we have $\boldsymbol{u}_0 + \lambda(\boldsymbol{v}_0 - \boldsymbol{u}_0) \in B_\delta(0)$. Now by the mean value theorem,

$$W(T,0;\boldsymbol{v}_0) - W(T,0;\boldsymbol{u}_0) = \int_0^1 (DW)(T,0;\boldsymbol{u}_0 + \lambda(\boldsymbol{v}_0 - \boldsymbol{u}_0))(\boldsymbol{v}_0 - \boldsymbol{u}_0)d\lambda.$$

Consider the Riemann sums

$$W_N = \sum_{j=1}^{N}(DW)(T,0;\boldsymbol{u}_0 + \lambda_j(\boldsymbol{v}_0 - \boldsymbol{u}_0))(\boldsymbol{v}_0 - \boldsymbol{u}_0)(\lambda_{j+1} - \lambda_j).$$

If we set $\lambda_{j+1} - \lambda_j = \beta_j$ and $\boldsymbol{v}_j = \boldsymbol{u}_0 + \lambda_j(\boldsymbol{v}_0 - \boldsymbol{u}_0)$ then $\sum_{j=1}^{N}\beta_j = 1$ and $\boldsymbol{v}_j \in B_\delta(0)$. We can then identify $W_N = L(\boldsymbol{v}_0 - \boldsymbol{u}_0)$ and hence from the part (i) of the lemma $W_N \in \overset{\circ}{\mathcal{K}}, \forall N$. Moreover the estimate (6.12) gives:

$$\|W_N\|_{D(A^\alpha)} \geq (1 + \eta - \epsilon)\|\boldsymbol{v}_0 - \boldsymbol{u}_0\|_{D(A^\alpha)}.$$

We now take the limit of these sums as $N \to \infty$ and obtain

$$W_N \to \{W(T,0;\boldsymbol{v}_0) - W(T,0;\boldsymbol{u}_0)\} \text{ as } N \to \infty$$

in the norm of $D(A^\alpha)$. Since $\mathcal{K}$ is closed we get $W(T,0;\boldsymbol{v}_0) - W(T,0;\boldsymbol{u}_0) \in \mathcal{K}$ and

$$\|W(T,0;\boldsymbol{v}_0) - W(T,0;\boldsymbol{u}_0)\|_{D(A^\alpha)} \geq (1 + \eta - \epsilon)\|\boldsymbol{v}_0 - \boldsymbol{u}_0\|_{D(A^\alpha)} \tag{6.13}$$

Hence $W(T, 0; \boldsymbol{v}_0) - W(T, 0; \boldsymbol{u}_0) \in \dot{\mathcal{K}}$.

Applying this results inductively we get,

$$\forall n \geq 1 \ , W(nT, 0; \boldsymbol{v}_0) - W(nT, 0; \boldsymbol{u}_0) \in \dot{\mathcal{K}}$$

and

$$\|W(nT, 0; \boldsymbol{v}_0) - W(nT, 0; \boldsymbol{u}_0)\|_{D(A^\alpha)} \geq (1 + \eta - \epsilon)^n \|\boldsymbol{v}_0 - \boldsymbol{u}_0\|_{D(A^\alpha)},$$

if $W(kT, 0; \boldsymbol{v}_0) - W(kT, 0; \boldsymbol{u}_0) \in B_\delta(0)$ holds for $k = 1, 2, \cdots, n - 1$. Since $1 + \eta - \epsilon > 1$, there exists an $m \in N$ such that $W(kT, 0; \boldsymbol{u}_0) \in \dot{\mathcal{K}}$ and $W(kT, 0; \boldsymbol{u}_0) \in B_\delta(0)$ cannot hold for all $k, 1 \leq k \leq m$. This proves part (iii) of the lemma as well as the part (ii) of the main theorem.

♣

Chapter 7

Invariant Manifold Theorem

Let us now refine the theory developed in the previous chapter under additional hypothesis regarding the spectrum of the monodromy operator. We will establish in particular the hyperbolicity property of each periodic (basic flow) orbit. The theory of hyperbolic sets for general C^1-maps is developed in [36, 93]. Such results for certain class of evolution equations can be found in [8] The analysis in this chapter improves the previous work [49] in two main aspects. Firstly the invariant manifold theorems are established independent of a specific topology. Secondly the analyticity of these manifolds are proved. In the language of aero/hydro dynamics, hyperbolicity property provides the *active* and *slave modes* of the orbit under study with respect to the basic periodic orbit. The computational implications of this result are evident. In fact invariant manifold theory has its origins in the early works in nonlinear hydrodynamic stability theory by Landau [51] and others [84, 92, 63]. *One should note that the central theorem 7.1 proved in this chapter provides the precise link between hydrodynamic transition and recent theories of finite dimensional dynamic systems[77, 71].*

Let the spectrum of the monodromy operator splits into two sets such that $\sigma(Z(T,0)) = \sigma_u \cup \sigma_s$ with $\sigma_u \cap \sigma_s = \text{empty}$(figure 7.1) and

$$b_s = \sup_{\lambda \in \sigma_s} |\lambda| < \inf_{\lambda \in \sigma_u} |\lambda| = b_u^{-1}. \tag{7.1}$$

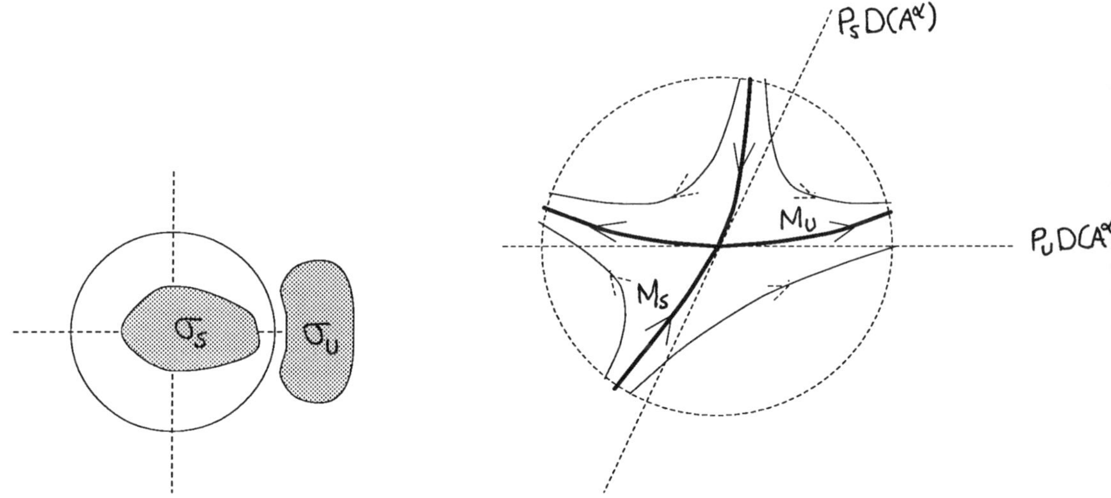

Figure 7.1 Spectrum And The Invariant Manifolds

Let $\boldsymbol{P_s}, \boldsymbol{P_u}$ be the corresponding projection operators defined as before.

Theorem 7.1 (The invariant manifold theorem)

Let $b_s, b_u < 1$. Then in a neighborhood $B_r(0) \subset \boldsymbol{D}(A^\alpha), \alpha \in [1/2, 1]$, there exists two unique, analytic manifolds M_s and M_u which are respectively the graphs of the maps, $\phi_s : \boldsymbol{P_s}\boldsymbol{D}(A^\alpha) \to \boldsymbol{P_u}\boldsymbol{D}(A^\alpha)$ and $\phi_u : \boldsymbol{P_u}\boldsymbol{D}(A^\alpha) \to \boldsymbol{P_s}\boldsymbol{D}(A^\alpha)$. The maps ϕ_s and ϕ_u are analytic with,

(i) $\phi_u(0) = \phi_s(0) = 0$,

$(ii) D\phi_u(0) = D\phi_s(0) = 0 \cdots$ tangency condition

(iii) manifolds M_s and M_u are locally invariant under the solution map $W(T, 0; \cdot)$:

$$W(T, 0; M_u \cap B_r(0)) \subset M_u \text{ and}$$

$$W(T, 0; M_s \cap B_r(0)) \subset M_s.$$

(iv) Stable manifold M_s satisfies

$$M_s \cap B_r(0) = \{u \in B_r(0) \text{ such that } \forall n \geq 0, W(nT, 0; u) \in B_r(0) \text{ and } \to 0 \text{ as } n \to \infty\}$$

$$\text{and } M_s \cap B_r(0) \cap \dot{\mathcal{K}} = empty \cdots (\mathcal{K} \text{ is the double cone defined by } (6.4))$$

Unstable manifold M_u satisfies,

$$M_u \cap B_\tau(0) = \{u \in B_\tau(0) \text{ such that } W(T,0;\cdot)^n u \text{ is defined } \forall n < 0$$

$$\text{and tends to zero as } n \to -\infty\} \text{ and } M_u \cap B_\tau(0) \subset \mathcal{K}.$$

(v) if $u \notin M_s$ then there exists $\delta > 0$ and $p \in N$ such that, $\|W(pT,0;u)\|_{D(A^\alpha)} > \delta$.

(vi) $\mathrm{dist}(M_u, W(T,0;u)) < \mathrm{dist}(M_u, u)$ for $u \in B_\tau(0) \cdots$ exponential attractive property of the unstable manifold.

$\mathrm{dist}(M_s, W(T,0;u)) > \mathrm{dist}(M_s, u)$ for $u \in B_\tau(0) \cdots$ **repelling property of the stable manifold.**

Proof:

let us first establish a result concerning F-analytic maps.

Lemma 7.1 *Let $\boldsymbol{H}_1, \boldsymbol{H}_2$ be Hilbert spaces over complex fields and $B_\tau(0) \subset \boldsymbol{H}_1$ be a closed ball of radius τ centered at the origin. Let $\boldsymbol{H}_{A\tau}$ be the space of F-analytic maps defined as,*

$$\boldsymbol{H}_{A\tau} = \{\phi : B_\tau(0) \subset \boldsymbol{H}_1 \to \boldsymbol{H}_2; \|\phi(\boldsymbol{x})\|_{H_2} \leq \tau, \forall \boldsymbol{x} \in B_\tau(0) \text{ and } \phi \text{ is F-analytic}\}.$$

Then $\boldsymbol{H}_{A\tau}$ is complete with the metric

$$dist(\phi, \phi') = \|\phi - \phi'\|_{H_{A\tau}} = \sup_{\boldsymbol{x} \in B_\tau(0)} \|\phi(\boldsymbol{x}) - \phi'(\boldsymbol{x})\|_{H_2}.$$

Moreover the subspace $\boldsymbol{H}_{A\tau k}$ defined as,

$$\boldsymbol{H}_{A\tau k} = \{\phi \in \boldsymbol{H}_{A\tau}; \phi(0) = 0; D\phi(0) = 0; \cdots D^k \phi(0) = 0\} \ , k \geq 0$$

is closed in $\boldsymbol{H}_{A\tau}$.

Proof:

Let ϕ_k be a Cauchy sequence in the norm of $\boldsymbol{H}_{A\tau}$ and $\phi_k \to \phi$. Since the maps ϕ_k are F-analytic in $B_\tau(0) \subset \boldsymbol{H}_1$, we have the map $\beta \to \phi_k(\boldsymbol{x} + \beta h)$ is $\boldsymbol{H}_2$-holomophic from $\mathcal{U}_\beta \subset C$ into $\boldsymbol{H}_2$. Here $\mathcal{U}_\beta$ is a complex neighborhood of the origin such that for fixed $\boldsymbol{x} \in B_\tau(0) \subset \boldsymbol{H}_1$, and $h \in \boldsymbol{H}_1$ we have $\boldsymbol{x} + \beta h \in B_\tau(0) \subset \boldsymbol{H}_1$. We have

$$\lim_{k \to \infty} \sup_{\boldsymbol{y} \in B_\tau(0)} \|\phi_k(\boldsymbol{y}) - \phi(\boldsymbol{y})\|_{H_2} = 0.$$

This implies,

$$\lim_{k \to \infty} \|\phi_k(x + \beta h) - \phi(x + \beta h)\|_{H_2} = 0 \text{ and}$$

$$\|\phi(x + \beta h)\|_{H_2} \le \tau, \forall \beta \in \mathcal{U}_\beta$$

with $x \in B_\tau(0)$, $h \in H_1$ both fixed such that $x + \beta h \in B_\tau(0)$.

We thus conclude by the vector valued generalization of the Vitali theorem [35] that the map $\beta \to \phi(x + \beta h)$ is H_2-holomorphic from $\mathcal{U}_\beta \to H_2$. Moreover the Cauchy integral,

$$\phi(x) = \frac{1}{2\pi i} \int_\Gamma \phi(x + \theta h) \frac{1}{\theta} d\theta$$

holds with Γ a closed circuit in $\mathcal{U}_\beta$ enclosing the origin. We write the G-derivative of ϕ as

$$\phi^{(1)}(x)h = \frac{d}{d\beta}\phi(x + \beta h)|_{\beta=0} = \frac{1}{2\pi i} \int_\Gamma \phi(x + \theta h) \frac{1}{\theta^2} d\theta,$$

$$\forall x \in B_\tau(0), \text{ and } \forall h \in H_1.$$

Now using the fact that $\|\phi(x)\|_{H_2}$ for $x \in B_\tau(0)$ is uniformly bounded by τ we get,

$$\|\phi^{(1)}(x)\|_{\mathcal{L}(H_1, H_2)} \le 1, \qquad \forall x \in B_\tau(0).$$

Thus the map ϕ is F-differentiable with the Frechet derivative

$$D\phi(x) = \phi^{(1)}(x) \in \mathcal{L}(H_1, H_2), \forall x \in B_\tau(0).$$

This in combination with the fact that ϕ is uniformly bounded in $B_\tau(0)$ gives us (from the definition of F-analyticity in ref[35]) that ϕ is F-analytic in $B_\tau(0)$. We also note that the convergence, $\lim_{k \to \infty} \phi_k(x + \beta h) = \phi(x + \beta h)$ implies that

$$\lim_{k \to \infty} \frac{\partial^n}{\partial \beta^n} \phi_k(x + \beta h) = \frac{\partial^n}{\partial \beta^n} \phi(x + \beta h), \qquad \forall n.$$

In particular this holds for $\beta = 0$ and $x = 0$:

$$\lim_{k \to \infty} D^n \phi_k(0) = D^n \phi(0), \qquad \forall n.$$

Hence we conclude that the space $H_{A_\tau k}$ is closed in the norm of H_{A_τ}.

We now return to the invariant manifold theorem and note that due to the proposition 5.1

$$\|P_s Z(T,0)\|_{\mathcal{L}(D(A^\alpha),D(A^\alpha))} = b_s < 1 \quad \text{and} \quad \|[P_u Z(T,0)]^{-1}\|_{\mathcal{L}(D(A^\alpha),D(A^\alpha))} = b_u < 1. \qquad (7.2)$$

We will denote by $\boldsymbol{H}^u_{A\tau k}$ the space obtained by setting (in the lemma 7.1) $\boldsymbol{H}_1 := P_u D(A^\alpha)$, $\boldsymbol{H}_2 := P_s D(A^\alpha)$ and $B_\tau(0) := B_{u\tau}(0)$. We will define a map $F_\phi : B_{u\tau}(0) \subset P_u D(A^\alpha) \to P_u D(A^\alpha)$ by

$$F_\phi = P_u W(T,0;\operatorname{graph}\phi), \qquad \phi \in \boldsymbol{H}^u_{A\tau 0}. \qquad (7.3)$$

That is for $x \in B_{u\tau}(0)$ we have $F_\phi(x) = P_u W(T,0;x+\phi(x))$.

Lemma 7.2 *For small enough τ the map F_ϕ is bi-Frechet analytic (both F_ϕ and F_ϕ^{-1} are analytic) homeomorphism. Moreover, $F_\phi^{-1}(B_{u\tau}(0)) \subset B_{u\tau}(0)$.*

Proof:
Note that for $x \in B_{u\tau}(0) \subset P_u D(A^\alpha)$ we can formally write

$$DF_\phi(x) = [DP_u W(T,0;x+\phi(x))] \circ (I + D\phi(x)).$$

The map $P_u W(T,0;\cdot) : B_\delta(0) \subset D(A^\alpha) \to P_u D(A^\alpha)$ is F-analytic for some $\delta > 0$. This means that $P_u W(T,0;y)$ is uniformly bounded for $y \in B_\delta(0)$ and that the Frechet derivative

$$DP_u W(T,0;y) \in \mathcal{L}(D(A^\alpha), P_u D(A^\alpha)), \qquad \forall y \in B_\delta(0) \subset D(A^\alpha).$$

Thus, noting the fact that, $x \in B_{u\tau}(0) \subset P_u D(A^\alpha)$ implies, $\|x+\phi(x)\|_{D(A^\alpha)} \leq 2\tau$ for $\phi \in \boldsymbol{H}_{A\tau 0}$, we get

$$DP_u W(T,0;x+\phi(x)) \in \mathcal{L}(D(A^\alpha), P_u D(A^\alpha)) \quad \text{for } 2\tau \leq \delta.$$

Moreover,

$$D\phi(x) \in \mathcal{L}(P_u D(A^\alpha), P_s D(A^\alpha)) \text{ for } x \in B_{u\tau}(0) \subset P_u D(A^\alpha)$$

due to the F-analyticity of ϕ. Hence

$$DF_\phi(x) \in \mathcal{L}(P_u D(A^\alpha), P_u D(A^\alpha)), \qquad \forall x \in B_{u\tau}(0)$$

This in combination with the fact that $F_\phi(\cdot)$ is uniformly bounded in $B_{u\tau}(0)$ (since $P_u W(T,0;\cdot)$ is uniformly bounded for $2\tau \leq \delta$) gives us the F-analyticity of the map F_ϕ in this ball.

Let us now establish the injectivity of this map. Note that the map $\psi(\cdot) = W(T,0;\cdot) - Z(T,0)$ is F-analytic in $B_\delta(0) \subset D(A^\alpha)$ with $\psi(0) = D\psi(0) = 0$. Thus the power series,

$$\psi(y) = \sum_{n \geq 2} \mathcal{H}_n(y, \cdots, y; T) \qquad \text{converges for } \|y\|_{D(A^\alpha)} \leq \delta, \tag{7.4}$$

where the n-linear maps

$$\mathcal{H}_n(\cdots; T) : B_\delta(0)^{\otimes n} \to D(A^\alpha) \qquad \text{are continuous.}$$

This also gives us

$$\|\psi(y) - \psi(y')\|_{D(A^\alpha)} \leq \eta \|y - y'\|_{D(A^\alpha)} \text{ for } y, y' \in B_\delta(0) \subset D(A^\alpha)$$

where η depends on $\|y\|, \|y'\|$ and $\to 0$ as $\sup(\|y\|, \|y'\|) \to 0$.

Similarly $\phi \in \boldsymbol{H}_{A\tau 0}$ has a power series

$$\phi(x) = \sum_{n \geq 1} \mathcal{A}_n(x, \cdots, x) \qquad \text{converges for } \|x\|_{D(A^\alpha)} \leq \tau. \tag{7.5}$$

Here the n-linear maps

$$\mathcal{A}_n(\cdots) : B_{u\tau}(0)^{\otimes n} \to P_s D(A^\alpha) \qquad \text{are continuous.}$$

Hence we again obtain the locally Lipschitz behaviour

$$\|\phi(x) - \phi(x')\|_{D(A^\alpha)} \leq \gamma \|x - x'\|_{D(A^\alpha)} \qquad \text{for } x, x' \in B_{u\tau}(0) \subset P_u D(A^\alpha), \gamma > 0.$$

Let us now consider,

$$F_\phi(x) - F_\phi(x') = P_u Z(T,0)(x - x') + P_u \psi(x + \phi(x))$$

118

$$-P_u\psi(x' + \phi(x')) \quad \text{for} \quad x, x' \in B_{u\tau}(0).$$

This gives

$$\|F_\phi(x) - F_\phi(x')\|_{D(A^\alpha)} \geq (b_u^{-1} - \eta(1 + \gamma))\|x - x'\|_{D(A^\alpha)} \tag{7.6}$$

and thus $F_\phi(\cdot)$ is injective locally. Noting that $F_\phi(0) = 0$, we get,

$$\|F_\phi(x)\|_{D(A^\alpha)} \geq \|x\|_{D(A^\alpha)} \quad \text{for small enough} \quad \tau.$$

In fact writing,

$$x = [P_u Z(T,0)]^{-1} F_\phi(x) - [P_u Z(T,0)]^{-1} P_u \psi(x + \phi(x))$$

and observing the fact that the map $[P_u Z(T,0)]^{-1} P_u \psi(\cdot + \phi(\cdot))$ is a contraction (since $b_u \eta(1+\gamma) < 1$) for small enough τ, we deduce that $F_\phi(\cdot)$ is onto and hence a homeomorphism. The estimate (7.6) (with $x' = 0$) gives $F_\phi^{-1}(B_{u\tau}(0)) \subset B_{u\tau}(0)$. This establishes the uniform boundedness of $F_\phi(\cdot)$ in this neighborhood. In order to prove the F-analyticity we need to show that F_ϕ^{-1} is Frechet differentiable in this neighborhood. Let us take $\bar{x} \in B_{u\tau}(0) \subset P_u D(A^\alpha)$ and set $F_\phi^{-1}(\bar{x}) = x \in B_{u\tau}(0)$. Now,

$$[P_u Z(T,0)]^{-1} \bar{x} = F_\phi^{-1}(\bar{x}) + [P_u Z(T,0)]^{-1} P_u \psi \circ (I + \phi) \circ F_\phi^{-1}(\bar{x}).$$

Taking the F-derivative formally at $\bar{x}$ gives

$$[P_u Z(T,0)]^{-1} = DF_\phi^{-1}(\bar{x})$$

$$+[P_u Z(T,0)]^{-1} DP_u \psi(x + \phi(x)) \circ (I + D\phi(x)) \circ DF_\phi^{-1}(\bar{x}).$$

Now note that the F-analyticity of the map $\psi(\cdot) : B_\delta(0) \subset D(A^\alpha) \to D(A^\alpha)$ implies the F-analyticity of its Frechet derivative

$$D\psi(\cdot)h : B_\delta(0) \subset D(A^\alpha) \to D(A^\alpha) \quad \text{for fixed } h \in D(A^\alpha).$$

This is due to a theorem on the Frechet analytic maps Ref [35], page 767. Thus the map $D\psi(\cdot)h$ is also locally Lipschitz in this neighborhood with some Lipschitz coefficient $\bar{\eta}$. Hence noting that $DP_u\psi(0) = 0$ we get

$$\|DP_u\psi(x + \phi(x))\|_{\mathcal{L}(D(A^\alpha), P_u D(A^\alpha))} \leq \bar{\eta}(1 + \gamma)\|x\|_{D(A^\alpha)} \quad \text{for} \quad x \in B_{u\tau}(0) \text{ and } 2\tau < \delta.$$

We have $D\phi(x) \in \mathcal{L}(P_u D(A^\alpha), P_s D(A^\alpha))$ for $x \in B_{u\tau}(0)$ due to the F-analyticity of ϕ and hence $I + D\phi(x) \in \mathcal{L}(P_u D(A^\alpha), D(A^\alpha))$. Thus the map

$$K = [P_u Z(T,0)]^{-1} D P_u \psi(x + \phi(x)) \circ (I + D\phi(x))$$

is a contraction for small enough τ because

$$\|K\|_{\mathcal{L}(P_u D(A^\alpha), P_u D(A^\alpha))} \leq b_u \bar{\eta}(1 + \gamma)\tau(1 + \|D\phi\|_{\mathcal{L}(P_u D(A^\alpha), P_s D(A^\alpha))}).$$

This gives $[I + K]^{-1} \in \mathcal{L}(P_u D(A^\alpha), P_u D(A^\alpha))$. We thus write

$$DF_\phi^{-1}(\bar{x}) = \{I + [P_u Z(t,0)]^{-1} D P_u \psi(x + \phi(x)) \circ (I + D\phi(x))\}^{-1} [P_u Z(t,0)]^{-1}$$

and $DF_\phi^{-1}(\bar{x}) \in \mathcal{L}(P_u D(A^\alpha), P_u D(A^\alpha))$ for $\bar{x} \in B_{u\tau}(0)$. In fact

$$\|DF_\phi^{-1}(\bar{x})\|_{\mathcal{L}(P_u D(A^\alpha), P_u D(A^\alpha))} \leq b_u + \circ(\tau).$$

This result, in combination with the fact that $F_\phi^{-1}(\cdot)$ is uniformly bounded in $B_{u\tau}(0)$ establishes the F-analyticity of this map .

♣

Let us now proceed with the construction of the unstable manifold. We define the **graph transform**

$$T_u \phi := P_s W(T, 0; \mathrm{graph}\phi), \qquad \phi \in H_{A\tau}^u \tag{7.7}$$

That is for $\bar{x} \in B_{u\tau}(0) \subset P_u D(A^\alpha)$ with $x = F_\phi^{-1}(\bar{x})$, we have

$$T_u \phi(\bar{x}) = P_s W(T, 0; x + \phi(x)).$$

Note that if ϕ is a fixed point of this transformation then $T_u \phi(\bar{x}) = \phi(\bar{x})$. This implies that

$$\{\bar{x}, \phi(\bar{x})\} = \{F_\phi(x), T_u \phi(\bar{x})\}$$

$$= \{P_u W(T, 0; x + \phi(x)), P_s W(T, 0; x + \phi(x))\} \subset P_u D(A^\alpha) \times P_s D(A^\alpha).$$

Thus $W(T, 0; \cdot) : \mathrm{graph}\ \phi \to \mathrm{graph}\ \phi$ in $B_{u\tau}(0)$. This means that the manifold characterized by graph ϕ is invariant to the action of the map $W(T, 0; \cdot)$. We will now establish such a fixed point for the graph transform T_u in $H_{A\tau1}^u$.

Lemma 7.3 *For small enough τ, the graph transform $T_u : H^u_{A\tau 1} \to H^u_{A\tau 1}$. Moreover it is a contraction in $H^u_{A\tau}$.*

Proof

First note that $\phi \in H^u_{A\tau 1}$ means $\phi(0) = D\phi(0) = 0$. Now,

$$T_u\phi(0) = P_s W(T, 0; 0 + \phi(0)) = 0$$

since $F_\phi^{-1}(0) = 0$. Writing

$$T_u\phi(\bar{x}) = P_s Z(T, 0)\phi \circ F_\phi^{-1}(\bar{x}) + P_s\psi \circ (I + \phi) \circ F_\phi^{-1}(\bar{x})$$

and taking F-derivative formally,

$$D(T_u\phi)(\bar{x}) = P_s Z(T, 0)D\phi(x) \circ DF_\phi^{-1}(\bar{x})$$

$$+ P_s D\psi(x + \phi(x)) \circ (I + D\phi(x)) \circ DF_\phi^{-1}(\bar{x}).$$

We note immediately that $D(T_u\phi)(0) = 0$. We will now establish the F-analyticity of the map $T_u\phi$. Firstly, for $\bar{x} \in B_{u\tau}(0)$,

$$\|T_u\phi(\bar{x})\|_{D(A^\alpha)} \leq b_s\|\phi(x)\|_{D(A^\alpha)} + \eta\|x + \phi(x)\|_{D(A^\alpha)}.$$

For $\bar{x} \in B_{u\tau}(0)$, we have $x \in B_{u\tau}(0)$ and hence $\|\phi(x)\|_{D(A^\alpha)} \leq \tau$. Thus

$$\|T_u\phi(\bar{x})\|_{D(A^\alpha)} \leq (b_s + 2\eta)\tau \leq \tau \quad \text{for small enough } \tau. \tag{7.8}$$

Moreover, the F-analyticity of the maps ϕ, $F_\phi(\cdot)$ and ψ imply that in these neighborhoods

$$D\phi(x) \in \mathcal{L}(P_u D(A^\alpha), P_s D(A^\alpha)),$$

$$DF_\phi^{-1}(\bar{x}) \in \mathcal{L}(P_u D(A^\alpha), P_u D(A^\alpha)),$$

and

$$P_s D\psi(x + \phi(x)) \in \mathcal{L}(D(A^\alpha), P_s D(A^\alpha))$$

if $2\tau \leq \delta$. Thus we get $DT_u\phi(\bar{x}) \in \mathcal{L}(P_u D(A^\alpha), P_s D(A^\alpha))$ for $\bar{x} \in B_{u\tau}(0)$. This result in combination with the uniform boundedness of $T_u\phi(\cdot)$ established in equation (7.8) implies the F-analyticity of $T_u\phi$. Hence we conclude that $T_u : H^u_{A\tau 1} \to H^u_{A\tau 1}$ for small enough τ.

We will now show that T_u is a contraction in the space $H^u_{A\tau}$. Let $\phi, \phi' \in H^u_{A\tau}$ and consider $\bar{x} \in B_{u\tau}(0)$. We denote $x = F_\phi^{-1}(\bar{x})$ and $x' = F_{\phi'}^{-1}(\bar{x})$. We note here that the unique invertability of the map F_ϕ does not require the condition $\phi(0) = 0$. Now,

$$\|T_u\phi(\bar{x}) - T_u\phi'(\bar{x})\|_{D(A^\alpha)} \leq b_s\|\phi(x) - \phi'(x')\|_{D(A^\alpha)}$$

$$+\eta\|\phi(x) - \phi'(x')\|_{D(A^\alpha)} + \eta\|x - x'\|_{D(A^\alpha)}$$

$$\leq (b_s + \eta)\|\phi(x) - \phi'(x)\|_{D(A^\alpha)} + (\eta + \gamma(\eta + b_s))\|x - x'\|_{D(A^\alpha)} \tag{7.9}$$

we also have $F_\phi(x) - F_{\phi'}(x') = 0$. Estimating this equation gives

$$\|x - x'\|_{D(A^\alpha)} \leq \frac{b_u\eta}{1 - b_u\eta(1 + \gamma)}\|\phi(x) - \phi'(x)\|_{D(A^\alpha)}. \tag{7.10}$$

Combining equations (7.9) and (7.10) gives

$$\|T_u\phi(\bar{x}) - T_u\phi'(\bar{x})\|_{D(A^\alpha)} \leq r\|\phi(x) - \phi'(x)\|_{D(A^\alpha)} \qquad \text{where} \tag{7.11}$$

$$r = b_s + \eta + b_u\eta\frac{(\eta + \gamma(\eta + b_s))}{1 - b_u\eta(1 + \gamma)} < 1 \quad \text{for small enough } \tau.$$

Hence $\|T_u\phi - T_u\phi'\|_{H^u_{A\tau}} \leq r\|\phi - \phi'\|_{H^u_{A\tau}}$ with $r < 1$.

♣

This lemma immediately gives us a unique fixed point ϕ_u (which is by construction an invariant manifold of $W(T, 0; \cdot)$ for the graph transform T_u with $\phi_u(0) = D\phi_u(0) = 0$. We will call this manifold the unstable manifold M_u.

Let us now establish the locally attractive property of M_u. Let us choose $y \in B_{\frac{\tau}{2}}(0) \subset D(A^\alpha)$ such that $W(T, 0; y) \in B_\tau(0) \subset D(A^\alpha)$. This is possible since the analyticity of the map implies

$$\|W(T, 0; y)\|_{D(A^\alpha)} \leq (\|Z(T, 0)\|_{\mathcal{L}(D(A^\alpha), D(A^\alpha))} + \eta)\|y\|_{D(A^\alpha)}.$$

let us write $y = x_u + x_s$ with $x_s = P_s y$ and $x_u = P_u y$.

Similarly write $\bar{y} = W(T, 0; y) = \bar{x}_s + \bar{x}_u$ with $\bar{x}_u = P_u\bar{y}$ and $\bar{x}_s = P_s\bar{y}$.

Lemma 7.4 *If $y \in B_{\frac{\tau}{2}}(0) \subset D(A^\alpha)$ be such that $\bar{y} = W(T,0;y) \in B_\tau(0)$ then*

$$\|\bar{x}_s - \phi_u(\bar{x}_u)\|_{D(A^\alpha)} \leq r\|x_s - \phi_u(x_u)\|_{D(A^\alpha)} \qquad \text{with } r < 1. \tag{7.12}$$

Proof:

Let us define a new map ϕ by

$$\phi(x'_u) = \phi_u(x'_u) + x_s - \phi_u(x_u) \qquad \text{for } x'_u \in B_{u\tau}(0) \subset P_u D(A^\alpha). \tag{7.13}$$

Note that $\phi(x_u) = x_s$ and that

$$F_\phi(x_u) = P_u W(T,0;x_u + \phi(x_u)) = P_u W(T,0;x_u + x_s) = \bar{x}_u.$$

Now,

$$T_u \phi(\bar{x}_u) = P_s W(T,0;x_u + \phi(x_u)) = P_s W(T,0;x_u + x_s) = P_s W(T,0;y)$$

That is $T_u \phi(\bar{x}_u) = \bar{x}_s$. Moreover ϕ_u is the fixed point for T_u and thus $T_u \phi_u(\bar{x}_u) = \phi_u(\bar{x}_u)$. Thus the inequality (7.12) in the lemma takes the form

$$\|T_u \phi(\bar{x}_u) - T_u \phi_u(\bar{x}_u)\|_{D(A^\alpha)} \leq r\|\phi(x_u) - \phi_u(x_u)\|_{D(A^\alpha)}.$$

Hence the lemma would be proved if we can show that $\phi \in H^u_{A\tau}$ (since T_u is a contraction in this class. See equation (7.11).). Note first that $D\phi(x'_u) = D\phi_u(x'_u) \in \mathcal{L}(P_u D(A^\alpha), P_s D(A^\alpha))$ for $x'_u \in B_{u\tau}(0)$ due to the F-analyticity of ϕ_u in this neighborhood. Moreover ϕ_u has the power series expansion

$$\phi_u(x'_u) = \sum_{n \geq 2} A_n(x'_u, \cdot, x'_u) \qquad \text{convergent in} \qquad B_{u\tau}(0).$$

This gives the Lipschitz property of ϕ_u with coefficient $\eta_1 \to 0$ as $\tau \to 0$. Thus

$$\|\phi(x'_u)\|_{D(A^\alpha)} \leq \|x_s\|_{D(A^\alpha)} + \eta_1 \|x'_u - x_u\|_{D(A^\alpha)}$$

$$\leq \frac{\tau}{2} + \eta_1 \frac{3\tau}{2} \leq \tau \qquad \text{for small enough} \quad \tau,$$

and hence $\phi \in H^u_{A\tau}$.

♣

This lemma establishes that the distance between a point and the manifold M_u decreases upon the action of the map $W(T, 0; \cdot)$ which proves the local attractivity of M_u.

Lemma 7.5 *The restriction of the solution map to the unstable manifold has a unique inverse locally. In other words*

$$W(T, 0; \cdot) : B_{\tau_1}(0) \cap M_u \to B_\tau(0) \cap M_u$$

is bijective for $B_{\tau_1}(0) \subset B_\tau(0) \subset D(A^\alpha)$

with τ_1 chosen sufficienty smaller than τ.

Proof:

Let us consider an element $\bar{y} \in M_u \cap B_\tau(0)$. We will find $y \in M_u \cap B_{\tau_1}(0)$ such that $\bar{y} = W(T, 0; y)$. We write $\bar{y} = \bar{x}_u + \bar{x}_s = \bar{x}_u + \phi_u(\bar{x}_u)$ and note that given $\bar{x}_u$, there exists unique $x_u = F_{\phi_u}^{-1}(\bar{x}_u)$. Thus we can find $y = x_s + x_u = \phi_u(x_u) + x_u$. That is $y = (I + \phi_u) \circ F_{\phi_u}^{-1}(P_u \bar{y})$. This shows that the map $W(T, 0; \cdot)$ is from $B_{\tau_1}(0) \cap M_u$ onto $B_\tau(0) \cap M_u$. To see the injectivity we use (7.6) to obtain the following estimate,

$$\|y - y'\|_{D(A^\alpha)} \leq (1 + \eta_1) \frac{b_u}{1 - b_u \eta(1 + \eta_1)} \|P_u \bar{y} - P_u \bar{y}'\|_{D(A^\alpha)}.$$

That is

$$\|y - y'\|_{D(A^\alpha)} \leq r \|\bar{y} - \bar{y}'\|_{D(A^\alpha)} \quad \text{with} \quad r < 1 \text{ for small enough } \tau.$$

We can thus write $y = W(T, 0; \cdot)^{-1} \bar{y}$ and get

$$\|W(T, 0; \cdot)^{-1} \bar{y} - W(T, 0; \cdot)^{-1} \bar{y}'\|_{D(A^\alpha)} \leq r \|\bar{y} - \bar{y}'\|_{D(A^\alpha)}.$$

♣

We also note that iterating the above steps for $\bar{y}' = 0$ give us

$$\|W(T, 0; \cdot)^{-p} \bar{y}\|_{D(A^\alpha)} \leq r^p \|\bar{y}\|_{D(A^\alpha)} \text{ with } r < 1$$

124

and hence taking $p \to \infty$ we get $W(T, 0; \cdot)^{-p} \bar{y} \to 0$ for $\bar{y} \in B_\tau(0) \cap M_u$. This completes the characterization of the unstable manifold.

Let us now construct the stable manifold. Let us define a space $H^s_{A\tau k}$ by setting (in the definition of the space $H_{A\tau k}$), $H_1 := P_s D(A^\alpha)$, $H_2 := P_u D(A^\alpha)$ and $B_\tau(0) := B_{s\tau}(0)$. Our goal is to construct an invariant manifold M_s given by the graph of $\phi \in H^s_{A\tau 1}$.

Let $x_s \in B_{s\tau}(0) \subset P_s D(A^\alpha)$ and define transforms T_s and G_ϕ as

$$T_s \phi(x_s) = [P_u Z(T, 0)]^{-1} \phi(\bar{x}_s) - [P_u Z(T, 0)]^{-1} P_u \psi(x_s + \phi(x_s)) \tag{7.14}$$

$$\bar{x}_s = G_\phi(x_s) = P_s W(T, 0; x_s + \phi(x_s)) \quad \text{with} \quad \phi \in H^s_{A\tau}. \tag{7.15}$$

We note that a fixed point ϕ_s of the map T_s satisfies $\phi_s(x_s) = T_s \phi_s(x_s)$ and this means

$$P_u Z(T, 0) \phi_s(x_s) + P_u \psi(x_s + \phi_s(x_s)) = \phi_s(\bar{x}_s).$$

That is $P_u W(T, 0; x_s + \phi_s(x_s)) = \phi_s(\bar{x}_s)$ with $\bar{x}_s = P_s W(T, 0; x_s + \phi_s(x_s))$. This implies $W(T, 0; \cdot) : \text{graph } \phi_s \to \text{graph } \phi_s$ and thus the graph of ϕ_s is an invariant manifold for the map $W(T, 0; \cdot)$. We will now establish the existence of such a fixed point for the transform T_s in the class $H^s_{A\tau 1}$.

Lemma 7.6 *Let $\phi \in H^s_{A\tau}$ then the map $G_\phi : B_{s\tau}(0) \subset P_s D(A^\alpha) \to P_s D(A^\alpha)$ is F-analytic and that $G_\phi(B_{s\tau}(0)) \subset B_{s\tau}(0)$ for small enough τ.*

Proof:
First note that for $x_s \in B_{s\tau}(0)$ and $\phi \in H^s_{A\tau}$ we have

$$\|G_\phi(x_s)\|_{D(A^\alpha)} \leq b_s \|x_s\|_{D(A^\alpha)} + \eta \|x_s + \phi(x_s)\|_{D(A^\alpha)}$$

$$\leq (b_s + 2\eta)\tau < \tau \quad \text{for small enough } \tau.$$

Moreover the Frechet derivative of $G_\phi(\cdot)$ is formally,

$$DG_\phi(x_s) = P_s Z(T, 0) + DP_s \psi(x_s + \phi(x_s)) \circ (I + D\phi(x_s)).$$

Note that if $2\tau < \delta$, then $\boldsymbol{x}_s + \boldsymbol{\phi}(\boldsymbol{x}_s) \in B_\delta(0) \subset D(A^\alpha)$ and hence due to the F-analyticity of the map $W(T, 0; \cdot)$

$$DP_s W(T, 0; \boldsymbol{x}_s + \boldsymbol{\phi}(\boldsymbol{x}_s)) \in \mathcal{L}(D(A^\alpha), P_s D(A^\alpha)).$$

Moreover, $D\boldsymbol{\phi}(\boldsymbol{x}_s) \in \mathcal{L}(P_s D(A^\alpha), P_u D(A^\alpha))$. Thus,

$$DG_\phi(\boldsymbol{x}_s) \in \mathcal{L}(P_s D(A^\alpha), P_s D(A^\alpha)) \quad \text{for} \quad \boldsymbol{x}_s \in B_{s\tau}(0).$$

Hence $G_\phi(\cdot)$ is F-analytic.

♣

Lemma 7.7 *For small enough τ the transform $T_s : H^s_{A\tau 1} \to H^s_{A\tau 1}$. Moreover it is a contraction in $H^s_{A\tau}$.*

Proof:

First note that

$$T_s\phi(0) = [P_u Z(T, 0)]^{-1}\phi \circ G_\phi(0) - [P_u Z(T, 0)]^{-1}P_u\psi(0 + \phi(0))$$

$$= 0 \quad \text{since} \quad G_\phi(0) = 0.$$

Also, taking the G-derivative of $T_s\phi(\cdot)$ at $\boldsymbol{x}_s$ gives

$$D[T_s\phi](\boldsymbol{x}_s) = [P_u Z(T, 0)]^{-1}D\phi(\bar{\boldsymbol{x}}_s) \circ DG_\phi(\boldsymbol{x}_s)$$

$$-[P_u Z(T, 0)]^{-1}DP_u\psi(\boldsymbol{x}_s + \phi(\boldsymbol{x}_s)) \circ (I + D\phi(\boldsymbol{x}_s)).$$

Since $D\psi(0) = 0$ and $D\phi(0) = 0$ we conclude that $D[T_s\phi](0) = 0$ for $\phi \in H^s_{A\tau 1}$.

Let us now establish the F-analyticity of the map $T_s\phi$. For $\boldsymbol{x}_s \in B_{s\tau}(0)$ and $\phi \in H^s_{A\tau}$ we have,

$$\|T_s\phi(\boldsymbol{x}_s)\|_{D(A^\alpha)} \leq b_u\tau + 2b_u\eta\tau \leq \tau \quad \text{for small enough } \tau.$$

Here we have used the fact that $\bar{\boldsymbol{x}}_s \in B_{s\tau}(0)$ and ψ is locally Lipschitz. Moreover, due to the F-analyticity of the maps ϕ, G_ϕ and ψ we get

$$[DT_s\phi](\boldsymbol{x}_s) \in \mathcal{L}(P_s D(A^\alpha), P_u D(A^\alpha)) \quad \text{for} \ \boldsymbol{x}_s \in B_{s\tau}(0) \ \text{and} \ 2\tau < \delta.$$

126

Hence, $T_s : H^s_{A\tau 1} \to H^s_{A\tau 1}$.

Let us now show that T_s is a contraction in $H^s_{A\tau}$. We have for $\phi, \phi' \in H^s_{A\tau}$ and $x_s \in B_{s\tau}(0) \subset P_s D(A^\alpha)$,

$$\|T_s\phi(x_s) - T_s\phi'(x_s)\|_{D(A^\alpha)} \le b_u \|\phi(\bar{x}_s) - \phi'(\bar{x}'_s)\|_{D(A^\alpha)}$$

$$+ b_u \eta \|\phi(x_s) - \phi'(x_s)\|_{D(A^\alpha)}. \qquad (7.16)$$

Here $\bar{x}_s = G_\phi(x_s)$ and $\bar{x}'_s = G_{\phi'}(x_s)$ both belong to $B_{s\tau}(0)$ due to the lemma proved earlier. Analyticity of ϕ implies local lipschitz property. Thus

$$\|\phi(\bar{x}_s) - \phi'(\bar{x}_s')\|_{D(A^\alpha)} \le \gamma_2 \|\bar{x}_s - \bar{x}'_s\|_{D(A^\alpha)} + \|\phi(\bar{x}'_s) - \phi'(\bar{x}'_s)\|_{D(A^\alpha)}$$

$$\le \gamma_2 \eta \|\phi(x_s) - \phi'(x_s)\|_{D(A^\alpha)} + \|\phi(\bar{x}'_s) - \phi'(\bar{x}'_s)\|_{D(A^\alpha)}. \qquad (7.17)$$

We combine equations (7.16) and (7.17) to get,

$$\|T_s\phi(x_s) - T_s\phi'(x_s)\|_{D(A^\alpha)} \le b_u \|\phi(\bar{x}'_s) - \phi'(\bar{x}'_s)\|_{D(A^\alpha)}$$

$$+ (1 + \gamma_2)\eta b_u \|\phi(x_s) - \phi'(x_s)\|_{D(A^\alpha)}.$$

That is

$$\|T_s\phi - T_s\phi'\|_{H^s_{A\tau}} \le b_u(1 + \eta(1 + \gamma_2))\|\phi - \phi'\|_{H^s_{A\tau}}.$$

This shows that for small enough τ , T_s is a contraction in $H^s_{A\tau}$.

♣

This result gives us the stable manifold M_s as the graph of ϕ_s. M_s is by construction invariant to the action of the solution map $W(T, 0; \cdot)$. Moreover the condition $\phi_s(0) = D\phi_s(0) = 0$ shows that the manifold is tangent to $P_s D(A^\alpha)$ at the origin.

Let us consider $x_s \in B_{s\tau_1}(0)$ and write $y = x_s + \phi(x_s) \in M_s \cap B_\tau(0) \subset D(A^\alpha), \tau > 2\tau_1$. The invariance of M_s gives

$$W(T, 0; y) = G_\phi(x_s) + \phi(\bar{x}_s) = \bar{x}_s + \phi(\bar{x}_s) \in M_s \cap B_\tau(0).$$

In fact we can show that $W(nT, 0; y) \in M_s \cap B_\tau(0), \forall n \ge 1$.

Note that for $y \in M_s \cap B_\tau(0)$ and $P_s y \in B_{s\tau_1}(0)$

$$\|P_s W(T, 0; y)\|_{D(A^\alpha)} \leq (b_s + \eta(1 + \gamma_2))\|P_s y\|_{D(A^\alpha)}.$$

Note that for small enough τ, $q = b_s + \eta(1 + \gamma_2) < 1$. Now iterating the above arguement we get,

$$\|P_s W(nT, 0; y)\|_{D(A^\alpha)} \leq q^n \|P_s y\|_{D(A^\alpha)}.$$

Note however that, the invariance of M_s implies,

$$W(nT, 0; y) = P_s W(nT, 0; y) + \phi_s(P_s W(nT, 0; y)).$$

Thus

$$\|W(nT, 0; y)\|_{D(A^\alpha)} \leq (1 + \gamma_2) q^n \|y\|_{D(A^\alpha)}.$$

This implies that on $M_s, W(nT, 0; \cdot) \to 0$ as $n \to \infty$.

Let us now consider a map $\Phi_M : B_\tau(0) \subset D(A^\alpha) \to D(A^\alpha)$ defined in the following way: For $x \in B_\tau(0), P_u x = x_u, P_s x = x_s$ and write

$$y_u = x_u - \phi_s(x_s) \in P_u D(A^\alpha)$$

$$y_s = x_s - \phi_u(x_u) \in P_s D(A^\alpha).$$

We then define the map $y = \Phi_M(x) = y_s + y_u$. We note that $\Phi_M : M_u \to P_u D(A^\alpha)$ and $\Phi_M : M_s \to P_s D(A^\alpha)$ in this neighborhood. The map Φ_M is actually a characterization of the distance from a point in this neighborhood to the manifolds M_s and M_u.

Lemma 7.8 *The map Φ_M is a bi-analytic homeomorphism between two neighborhoods of the origin of $D(A^\alpha)$ such that*

$$B_{\tau/2}(0) \subset \Phi_M(B_\tau(0)) \subset B_{2\tau}(0).$$

Proof:

Let us first establish the F-analyticity of the map Φ_M. Note that for $x \in B_\tau(0)$

$$\|\Phi_M(x)\|_{D(A^\alpha)} \leq \|x_s + x_u\|_{D(A^\alpha)} + \|\phi_s(x_s)\|_{D(A^\alpha)} + \|\phi_u(x_u)\|_{D(A^\alpha)}.$$

128

Since ϕ_s and ϕ_u are F-analytic maps with $\phi_s(0) = \phi_u(0) = D\phi_u(0) = D\phi_s(0) = 0$, we have

$$\|\phi_s(x_s)\|_{D(A^\alpha)} \leq \eta_2 \|x_s\|_{D(A^\alpha)} \qquad \text{and}$$

$$\|\phi_u(x_u)\|_{D(A^\alpha)} \leq \eta_1 \|x_u\|_{D(A^\alpha)}$$

with $\eta_1, \eta_2 \to 0$ as $\|x\|_{D(A^\alpha)} \to 0$. Now using the fact that $\|x_s\|_{D(A^\alpha)} + \|x_u\|_{D(A^\alpha)} = \|x\|_{D(A^\alpha)}$ for the equivalent norm,

$$\|\Phi_M(x)\|_{D(A^\alpha)} \leq (1 + \eta_0)\|x\|_{D(A^\alpha)} \quad \text{where} \quad \eta_0 = \sup(\eta_1, \eta_2).$$

This gives us $\Phi_M(B_\tau(0)) \subset B_{2\tau}(0)$ for small enough τ. In order to prove the analyticity we need to verify the Frechet differentiability of Φ_M in $B_\tau(0)$. Let $x \in B_\tau(0)$ and $h \in D(A^\alpha)$ both be fixed. Let us write $P_u h = h_u$ and $P_s h = h_s$. Now the G-derivative of $\Phi_M(\cdot)$ will be

$$\frac{\Phi_M(x + \beta h)}{d\beta}\Big|_{\beta=0} = \frac{d}{d\beta}[(x + \beta h) - \phi_s(x_s + \beta h_s) - \phi_u(x_u + \beta h_u)]\big|_{\beta=0}$$

That is

$$\Phi_M^{(1)}(x)h = h - D\phi_s(x_s)h_s - D\phi_u(x_u)h_u. \tag{7.18}$$

Since $\phi_s(\cdot)$ and $\phi_u(\cdot)$ are F-analytic maps, we get

$$\|\Phi_M^{(1)}(x)h\|_{D(A^\alpha)} \leq (1 + \|D\phi_s(x_s)\|_{\mathcal{L}(P_s D(A^\alpha), P_u D(A^\alpha))}$$

$$+ \|D\phi_u(x_u)\|_{\mathcal{L}(P_u D(A^\alpha), P_s D(A^\alpha))})\|h\|_{D(A^\alpha)}.$$

Denoting by η_s, η_u the bounds for the operator norms on the right hand side we get

$$\|\Phi_M^{(1)}(x)\|_{\mathcal{L}(D(A^\alpha), D(A^\alpha))} \leq 1 + \eta_s + \eta_u$$

Thus $\Phi_M(\cdot)$ is Frechet differentiable with the Frechet derivative $D\Phi_M(x) = \Phi_M^{(1)}(x) \in \mathcal{L}(D(A^\alpha); D(A^\alpha))$ for $x \in B_\tau(0)$. This gives the F-analyticity of the map Φ_M.

Note here that the analyticity of the maps ϕ_s and ϕ_u imply the analyticity of the Frechet derivatives

$$D\phi_s(\cdot)h_s : B_{s\tau}(0) \subset P_s D(A^\alpha) \to P_u D(A^\alpha) \quad \text{and}$$

$$D\phi_u(\cdot)h_u : B_{u\tau}(0) \subset P_u D(A^\alpha) \to P_s D(A^\alpha)$$

for fixed $h_s \in P_s D(A^\alpha)$ and $h_u \in P_u D(A^\alpha)$. This in combination with the fact that $D\phi_s(0) = D\phi_u(0) = 0$ imply that the maps $D\phi_u(\cdot)h_u$ and $D\phi_s(\cdot)h_s$ are locally Lipschitz and η_u, η_s defined above satisfy $\eta_s \to 0$ as $\|x_s\|_{D(A^\alpha)} \to 0$ and $\eta_u \to 0$ as $\|x_u\|_{D(A^\alpha)} \to 0$.

Let us now study the inverse of the map Φ_M. If we write $\Phi_M = I + S_M$, then $S_M : B_\tau(0) \to B_{2\tau}(0) \subset D(A^\alpha)$ is a contraction map. To see this take $x, x' \in B_\tau(0)$ and consider

$$\|S_M(x) - S_M(x')\|_{D(A^\alpha)} \le \eta_1 \|x_u - x'_u\|_{D(A^\alpha)} + \eta_2 \|x_s - x'_s\|_{D(A^\alpha)}$$

$$\le \eta_0 \|x - x'\|_{D(A^\alpha)}$$

with $\eta_0 \to 0$ as $\|x\|_{D(A^\alpha)}, \|x'\|_{D(A^\alpha)} \to 0$. Hence the map Φ_M is uniquely invertable in a neighborhood of the origin of $D(A^\alpha)$. Now for $x \in B_\tau(0)$

$$\|\Phi_M(x)\|_{D(A^\alpha)} \ge (1 - \eta_0)\|x\|_{D(A^\alpha)} \ge \frac{1}{2}\|x\|_{D(A^\alpha)}$$

for τ small enough. This gives us the result,

$$\Phi_M^{-1}(B_{\frac{\tau}{2}}(0)) \subset B_\tau(0) \subset D(A^\alpha).$$

To establish the F-analyticity of the map Φ_M^{-1} we need to prove the F-differentiability. For $y \in B_{\tau/2}(0)$ let us write $y = (I + S_M) \circ \Phi_M^{-1}(y)$ and take the G-derivative at y to get,

$$I = (I + DS_M(x)) \circ D\Phi_M^{-1}(y)$$

with $x = \Phi_M^{-1}(y) \in B_\tau(0)$. Note that for fixed $h \in D(A^\alpha)$

$$DS_M(x)h = -D\phi_s(x_s)h_s - D\phi_u(x_u)h_u \quad \text{and}$$

$$\|DS_M(x)\|_{\mathcal{L}(D(A^\alpha),D(A^\alpha))} \le \eta_s + \eta_u \to 0 \quad \text{as} \quad \|x\|_{D(A^\alpha)} \to 0.$$

Hence the linear operator $I + DS_M$ has continuous inverse and,

$$\|[I + DS_M(x)]^{-1}\|_{\mathcal{L}(D(A^\alpha),D(A^\alpha))} \le \frac{1}{1 + \eta_s + \eta_u}.$$

Thus

$$D\Phi_M^{-1}(y) \in \mathcal{L}(D(A^\alpha), D(A^\alpha)) \quad \text{for} \quad y \in B_{\tau/2}(0)$$

and this proves the F-analyticity of the map Φ_M^{-1} .

130

Let us define a map $\mathcal{G}(\cdot) : B_{\frac{\tau}{2}}(0) \subset D(A^\alpha) \to D(A^\alpha)$ by

$$\mathcal{G}(y) = \Phi_M \circ W(T, 0; \Phi_M^{-1}(y)). \tag{7.19}$$

Then

Lemma 7.9 $\mathcal{G}(\cdot)$ *is F-analytic and the linear manifolds* $P_s D(A^\alpha)$ *and* $P_u D(A^\alpha)$ *are locally invariant to the action of this map . Moreover*

$$(i) D\mathcal{G}(0) = Z(T, 0) \quad and$$

$$(ii) \|P_u \mathcal{G}(y)\|_{D(A^\alpha)} \geq q_1 \|P_u y\|_{D(A^\alpha)}, \forall y \in B_{\frac{\tau}{2}}(0) \backslash P_s D(A^\alpha)$$

with some $q_1 > 1$.

Proof:

$\mathcal{G}(\cdot)$ is defined as a composition of analytic maps and hence the F-analyticity follows. We have for $y \in B_{\tau/2}(0) \subset D(A^\alpha)$

$$D\mathcal{G}(y) = D\Phi_M(\bar{x}) \circ DW(T, 0; x) \circ D\Phi_M^{-1}(y)$$

with $x = \Phi_M^{-1}(y) \in B_\tau(0)$ and

$$\bar{x} = W(T, 0; x) \in B_{\tau_1}(0) \subset D(A^\alpha).$$

Set $y = 0$ and then $x = \bar{x} = 0$

$$D\mathcal{G}(0) = D\Phi_M(0) \circ Z(T, 0) \circ D\Phi_M^{-1}(0).$$

Since $D\Phi_M(0) = D\Phi_M^{-1}(0) = I$ we get $D\mathcal{G}(0) = Z(T, 0)$. Now note that if $y \in P_u D(A^\alpha)$ (this in turn corresponding to $\Phi_M^{-1}(y) \in M_u$) then $W(T, 0; \Phi_M^{-1}(y)) \in M_u$ due to the invariance of M_u. Since $\Phi_M : M_u \to P_u D(A^\alpha)$ we get $\mathcal{G}(\cdot) : P_u D(A^\alpha) \to P_u D(A^\alpha)$. A similar arguement gives $\mathcal{G}(\cdot) : P_s D(A^\alpha) \to P_s D(A^\alpha)$. This means

$$\forall y \in B_{\tau/2}(0) \subset D(A^\alpha), P_u \mathcal{G}(P_s y) = P_s \mathcal{G}(P_u y) = 0.$$

Moreover, writing $\mathcal{G}(y) = Z(T,0)y + \hat{\psi}(y)$ we get

$$P_s\hat{\psi}(P_u y) = P_u\hat{\psi}(P_s y) = 0.$$

We also observe the following properties of $\hat{\psi}(\cdot)$:

$$\hat{\psi}(0) = \mathcal{G}(0) = \Phi_M \circ W(T,0;\Phi_M^{-1}(0)) = 0$$

and

$$D\hat{\psi}(0) = D\mathcal{G}(0) - Z(T,0) = 0.$$

Combining this result with F-analyticity gives us local Lipschitz continuity of $\hat{\psi}$ with coefficient $\hat{\eta} \to 0$ as $\tau \to 0$. Let us now consider $x \in B_r(0)\backslash M_s$. This means $P_u y = y_u \neq 0$. Let us write

$$P_u\mathcal{G}(y) = P_u Z(T,0)y_u + P_u\hat{\psi}(y_u + y_s) - P_u\hat{\psi}(y_s).$$

This gives,

$$\|P_u\mathcal{G}(y)\|_{D(A^\alpha)} \geq b_u^{-1}\|y_u\|_{D(A^\alpha)} - \hat{\eta}\|y_u\|_{D(A^\alpha)}.$$

Note that for small enough τ, $q_1 = b_u^{-1} - \hat{\eta} > 1$. Thus

$$\|P_u\mathcal{G}(y)\|_{D(A^\alpha)} \geq q_1\|P_u y\|_{D(A^\alpha)}, \forall y \in B_{\frac{r}{2}}(0)\backslash P_s D(A^\alpha)$$

$$\clubsuit$$

Note also that for such a $y = \Phi_M^{-1}(x)$ we have $x \in B_r(0)\backslash M_s$ and that, the statement(ii) of the lemma 7.9 provides

$$\|P_u W(T,0;x) - \phi_s(P_s W(T,0;x))\|_{D(A^\alpha)} \geq q_1\|x_u - \phi_s(x_s)\|_{D(A^\alpha)}.$$

This implies that the points not on the stable manifold are mapped farther from this manifold under the action of the solution map $W(T,0;\cdot)$. This is the repelling property of M_s. Note also that, for $y \in B_{\frac{r}{2}}(0)\backslash P_s D(A^\alpha)$

$$\|\mathcal{G}^n(y)\|_{D(A^\alpha)} \geq \|P_u\mathcal{G}^n(y)\|_{D(A^\alpha)} \geq q_1^n\|P_u y\|_{D(A^\alpha)}$$

provided $\mathcal{G}^k(\boldsymbol{y}) \in B_{\frac{r}{2}}(0)$ for $k = 0, 1, \cdots, n - 1$. Since $q_1 > 1$, this cannot hold for all n. Thus there exists an n such that $\mathcal{G}^n(\boldsymbol{y})$ falls outside $B_{\frac{r}{2}}(0)$ and this means $W(nT, 0; \Phi_M^{-1}(\boldsymbol{y}))$ leaves $B_r(0)$. This completes the Proof of the invariant manifold theorem.

♣

Appendix A

Global Attractors For Two Dimensional Viscous Flows

This section provides a brief description of the global attractor concept advanced by Ladyzhenskaya [47],[46]. Further studies of this attractor can be found in [15],[86],[66],[41]. Let $\Omega \subset \boldsymbol{R}^2$ be an open *bounded* set with boundary $\partial\Omega$ be of class $\boldsymbol{C}^\infty$. Let $\boldsymbol{f} : \Omega \to \boldsymbol{R}^2$ be a given time independent vector field such that $\boldsymbol{f} \in \boldsymbol{H}$. We will analyze the long time behavior of the Navier-Stokes system

$$u_t + (\boldsymbol{u} \cdot \nabla)\boldsymbol{u} = -\nabla p + \nu \Delta \boldsymbol{u} + \boldsymbol{f} \text{ in } \Omega \times (0, \infty)$$

$$\nabla \cdot \boldsymbol{u} = 0 \text{ in } \Omega \times (0, \infty) \tag{A.1}$$

$$\boldsymbol{u}(x,t) = 0 \text{ for } (x,t) \in \partial\Omega \times [0, \infty)$$

$$\text{and } \boldsymbol{u}(x,0) = \boldsymbol{u}_0(\boldsymbol{x}), \text{ for } x \in \Omega.$$

Let us take $\boldsymbol{u}_0 \in \boldsymbol{H}$. Applying the orthogonal projector $\boldsymbol{P}_H : L^2(\Omega) \to \boldsymbol{H}$ to (A.1) gives us the evolution system

$$u_t + \nu Au + B(\boldsymbol{u}, \boldsymbol{u}) = \boldsymbol{f} \ \in \boldsymbol{H}, \ t > 0 \tag{A.2}$$

$$\boldsymbol{u}(0) = \boldsymbol{u}_0 \ \in \boldsymbol{H}.$$

For this system we have the following unique solvability result.

Theorem A.1 *For the system A.2 with arbitrary $\boldsymbol{u}_0, \boldsymbol{f} \in \boldsymbol{H}$, $\exists$ a unique mild solution*

$$\boldsymbol{u} \in L^2(0,T;\boldsymbol{V}) \cap C([0,T];\boldsymbol{H}), \quad \forall T > 0$$

The solution is holomorphic in a neighborhood of the positive time axis (figure A.1) and is analytic in the initial as well as forcing datum in neighborhoods in $\boldsymbol{H}$.

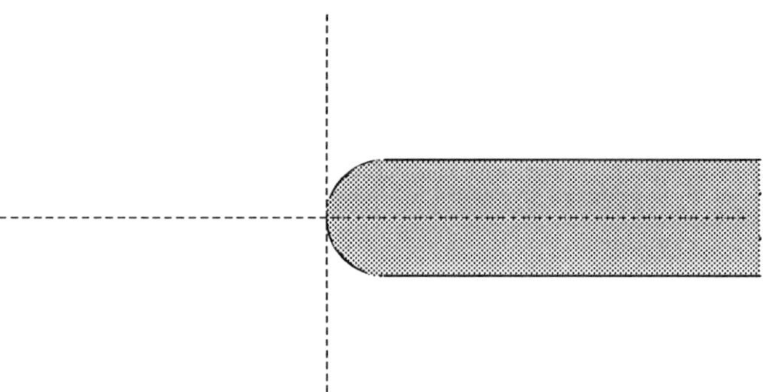

Figure A.1: Time Analyticity Domain

As remarked earlier the existence result is due to Hopf (1951) and the uniqueness theorem is due to Lions and Prodi (1959). The analyticity with respect to time can be established using semigroup methods [23],[40] or by complexified Galerkin methods [86]. Analyticity in initial data is proved in chapter 5. Analyticity with respect to forcing data can be deduced from a slightly more involved theorem in [82] concerning exterior hydrodynamics. The following result due to Prodi [68] establishes the continuity of the hydrodynamic semigroup in the weak topology.

Theorem A.2 *Let $W(t,0;\cdot) : \boldsymbol{H} \to \boldsymbol{H}$, $t \in (0,\infty)$ be the hydrodynamic semigroup associated with the evolution system (A.2). Suppose that $\{\boldsymbol{u}_0^{(i)}\} \in \boldsymbol{H}$ be a bounded sequence converging weakly to $\boldsymbol{u}_0 \in \boldsymbol{H}$. Then, $\forall t \in (0,\infty)$, the sequence $\{\boldsymbol{u}^{(i)}(t) = W(t,0;\boldsymbol{u}_0^{(i)})\} \in \boldsymbol{H}$ converges weakly to $\boldsymbol{u}(t) = W(t,0;\boldsymbol{u}_0) \in \boldsymbol{H}$.*

We will now derive certain global bounds for the solution. We can get a weak formulation from the system (A.2) by taking inner product in $\boldsymbol{H}$ with $\boldsymbol{w} \in \boldsymbol{V}$,

$$(\frac{\partial \boldsymbol{u}}{\partial t}, \boldsymbol{w})_{L^2(\Omega)} + \nu(\nabla \boldsymbol{u}, \nabla \boldsymbol{w})_{L^2(\Omega)} + b(\boldsymbol{u}, \boldsymbol{u}, \boldsymbol{w}) = (\boldsymbol{f}, \boldsymbol{w})_{L^2(\Omega)} \quad \forall \boldsymbol{w} \in \boldsymbol{V}. \quad (A.3)$$

We will first obtain the *energy estimate* by setting $w = u$ and using the fact that $b(u, u, u) = 0$,

$$\frac{1}{2}\frac{d}{dt}\|u\|^2_{L^2(\Omega)} + \nu\|\nabla u\|^2_{L^2(\Omega)} = (f, u)_{L^2(\Omega)} \leq \|f\|_{L^2(\Omega)}\|u\|_{L^2(\Omega)}.$$

Using the Poincaré's Lemma $\|\nabla u\|^2_{L^2(\Omega)} \geq \mu_1\|u\|^2_{L^2(\Omega)}$ and Young's inequality we get

$$\frac{d}{dt}\|u\|^2_{L^2(\Omega)} + 2\nu\|\nabla u\|^2_{L^2(\Omega)} \leq \frac{\|f\|^2_{L^2(\Omega)}}{\mu_1\nu} + \nu\|\nabla u\|^2_{L^2(\Omega)}.$$

That is

$$\frac{d}{dt}\|u\|^2_{L^2(\Omega)} + \nu\mu_1\|u\|^2_{L^2(\Omega)} \leq \frac{\|f\|^2_{L^2(\Omega)}}{\mu_1\nu}.$$

Here μ_1 is the smallest eigenvalue of Stokes operator A. We then set $\alpha = \nu\mu_1$ and $\rho_0 = \alpha^{-1}\|f\|_{L^2(\Omega)}$ and obtain using Gronwall's lemma[91],

$$\|u(t)\|^2_{L^2(\Omega)} \leq \|u(0)\|^2_{L^2(\Omega)}e^{-\alpha t} + \rho_0^2[1 - e^{-\alpha t}], \quad \alpha > 0. \tag{A.4}$$

We note that when $\|u(x, 0)\|_{L^2(\Omega)} > \rho_0$, the energy estimate (A.4) shows that $\|u(t)\|_{L^2(\Omega)}$ is bounded by a monotone decreasing exponential function. On the other hand, if

$$\|u(x, 0)\|_{L^2(\Omega)} \leq \rho_0, \text{ then } \|u(x, t)\|_{L^2(\Omega)} \leq \rho_0 \ \forall t \geq 0.$$

Hence for any ball

$$B_R(H) = \{u(0) \in H; \ \|u(0)\|_{L^2(\Omega)} \leq R\},$$

there is a ball $B_{r_0}(H)$ in H centered at origin with radius $r_0 > \rho_0$ such that

$$W(t, 0; B_R(H)) \subseteq B_{r_0}(H), \quad \text{for } t \geq t_0(B_R) = \frac{1}{\alpha}\ln\frac{R^2 - \rho_0^2}{r_0^2 - \rho_0^2}. \tag{A.5}$$

The ball $B_{r_0}(H)$ is said to be **exponentially absorbing** and **invariant** under the action of the map $W(t, 0; \cdot)$. *This result was first noted in a famous paper by Hopf in 1941* [37].

Let us now proceed to get other estimates. We will begin with the energy estimate

$$\frac{d}{dt}\|u\|^2_{L^2(\Omega)} + \nu\|\nabla u\|^2_{L^2(\Omega)} \leq \frac{\|f\|^2_{L^2(\Omega)}}{\mu_1\nu}.$$

Integrating from t to $t + 1$ we obtain

$$\|u(t + 1)\|^2_{L^2(\Omega)} + \nu\int_t^{t+1}\|\nabla u(\tau)\|^2_{L^2(\Omega)}d\tau \leq \frac{\|f\|^2_{L^2(\Omega)}}{\mu_1\nu} + \|u(t)\|^2_{L^2(\Omega)}.$$

Taking $u_0 \in B_R(H)$ and $t \geq t_0(B_R)$ we get,

$$\int_t^{t+1} \|\nabla u\|_{L^2(\Omega)}^2 d\tau \leq \frac{1}{\nu}\left[r_0^2 + \frac{1}{\mu_1\nu}\|f\|_{L^2(\Omega)}^2\right] = \sigma_0. \tag{A.6}$$

We now recall the uniform Gronwall inequality [22]:

Lemma A.1 Uniform Gronwall Inequality

Let g, h, y be three positive locally integrable functions for $t_0 \leq t < +\infty$ which satisfy

$$\frac{dy}{dt} \leq gy + h \qquad\qquad \forall\ t \geq t_0\ \text{and}$$

$$\int_t^{t+1} g(s)ds \leq \alpha_1\,, \qquad \int_t^{t+1} h(s)ds \leq \alpha_2\,, \qquad \int_t^{t+1} y(s)ds \leq \alpha_3\,,$$

for all $t \geq t_0$, where $\alpha_1, \alpha_2, \alpha_3$ are positive constants. Then

$$y(t+1) \leq (\alpha_2 + \alpha_3)exp(\alpha_1)\,, \qquad \forall\ t \geq t_0\,. \tag{A.7}$$

Proof : Integrating the differential inequality from τ to $t+1$,

$$y(t+1) \leq y(\tau) + \int_\tau^{t+1} h(\xi)d\xi + \int_\tau^{t+1} g(\xi)y(\xi)d\xi\,, \qquad \forall\ t \leq \tau \leq t+1\,.$$

Since the functions g, h and y are positive ,

$$y(t+1) \leq y(\tau) + \alpha_2 + \int_t^{t+1} g(\xi)y(\xi)d\xi\,.$$

Integrating again with respect to τ,

$$y(t+1) \leq (\alpha_2 + \alpha_3) + \int_t^{t+1} g(\xi)y(\xi)d\xi\,.$$

Now applying the Gronwall's inequality [91], we get

$$y(t+1) \leq (\alpha_2 + \alpha_3)exp(\alpha_1)\,.$$

♣

Let us now set $w = Au$ in (A.2),

$$\frac{1}{2}\frac{d}{dt}\|\nabla u\|^2_{L^2(\Omega)} + \nu\|Au\|^2_{L^2(\Omega)} \leq C_1\|u\|^{1/2}_{L^2(\Omega)}\|\nabla u\|_{L^2(\Omega)}\|Au\|^{3/2}_{L^2(\Omega)}$$

$$+ \|f\|_{L^2(\Omega)}\|Au\|_{L^2(\Omega)}. \tag{A.8}$$

In the above, we have used the following estimate for trilinear form $b(u, u, Au)$ which is valid only for $n = 2$ (see Temam [87] for example),

$$|b(u, u, Au)| \leq C_1\|u\|^{1/2}_{L^2(\Omega)}\|\nabla u\|_{L^2(\Omega)}\|Au\|^{3/2}_{L^2(\Omega)}.$$

Using the Young's inequality we get

$$\frac{d}{dt}\|\nabla u\|^2_{L^2(\Omega)} + 2\nu\|Au\|^2_{L^2(\Omega)} \leq 2\nu\|Au\|^2_{L^2(\Omega)} + C_2\|u\|^2_{L^2(\Omega)}\|\nabla u\|^4_{L^2(\Omega)} + \frac{1}{\nu}\|f\|^2_{L^2(\Omega)}$$

That is

$$\frac{d}{dt}\|\nabla u\|^2_{L^2(\Omega)} \leq C_2\|u\|^2_{L^2(\Omega)}\|\nabla u\|^4_{L^2(\Omega)} + \frac{\|f\|^2_{L^2(\Omega)}}{\nu}.$$

By applying the uniform Gronwall inequality (Lemma A.1) we get

$$\|\nabla u(t)\|^2_{L^2(\Omega)} \leq (\sigma_0 + \frac{\|f\|^2_{L^2(\Omega)}}{\nu})\,\exp(C_2\,{r_0}^2\,\sigma_0), \quad \text{for } t \geq t_0(B_R) + 1.$$

This means that there exists an absorbing ball $B_{r_1}(V)$ in V such that

$$W(t, 0; B_R(H)) \subseteq B_{r_1}(V), \quad \text{for } t \geq t_0(B_R) + 1.$$

Let us define the global attractor Λ as the ω-limit set of the absorbing ball $B_{r_1}(V)$. That is

$$\Lambda = \bigcap_{\tau \geq 0} cl\left(\bigcup_{t \geq \tau} W(t, 0; B_{r_1}(V))\right).$$

By the theorems A.1, A.2 and by the compactness of the embedding $V \subset H$ we can conclude that $W(t, 0; \cdot)$ maps a closed bounded set in H into compact set in H. Thus Λ is compact in H.

We will now prove the central result,

Theorem A.3 The Navier-Stokes system (A.2) defines a dynamical system on Λ. That is if $u_0 \in \Lambda$ then (A.2) is uniquely solvable in $B_{r_0}(H)$ for $t > 0$ as well as for $t < 0$.

Proof:

Let $B_{r_0}(t) = W(t, 0; B_{r_0}(H))$. Then (A.4) implies that $B_{r_0}(t) \subseteq B_{r_0}(H)$, $\forall t \geq 0$. Moreover, by theorem A.2, $B_{r_0}(t)$ is closed. Hence,

$$\Lambda = \bigcap_{t \geq 0} B_{r_0}(t)$$

Let us now suppose that $u(t) \in B_{r_0}(H), \forall t \in R$ and uniquely solves (A.2) with $u(0) = u_0 \in B_{r_0}(H)$. Then u_0 has to be an element of the global attractor. To see this we will set $u^{(1)}(t) = u(t - \tau) \in B_{r_0}(H)$. Then $u^{(1)}(t)$ solves (A.2) for $t \geq 0$ with $u^{(1)}(\tau) = u_0 \in B_{r_0}(\tau)$. This is valid for each $\tau \geq 0$ and hence we should have $u_0 \in B_{r_0}(\tau), \forall \tau \geq 0$. Thus

$$u_0 \in \bigcap_{\tau \geq 0} B_{r_0}(\tau) = \Lambda.$$

We will now suppose on the other hand that $u_0 \in \Lambda$. Since Λ is the $\omega - limit\ set$ of the ball $B_{r_0}(H)$ we have, for each $t_m = m,\ m = 1, 2, \cdots \exists\ u^{(m)}(t) \in B_{r_0}(H)$, which solves (A.2) for $t \geq 0$ and $u^{(m)}(t_m) = u_0$. Then, $\hat{u}^{(m)}(t) = u^{(m)}(t + t_m)$ solves (A.2) for $t \geq -t_m$. Note now that the solutions $\hat{u}^{(m)}(t)$ and $\hat{u}^{(m+1)}(t)$ are defined for $t \geq -t_m$ and $\hat{u}^{(m)}(0) = \hat{u}^{(m+1)}(0) = u_0$. However the time holomorphic property in theorem A.1 implies that the hydrodynamic semigroup is injective and hence we have the backward uniqueness result: if $u^{(m)}(t_m) = u^{(m+1)}(t_m)$ then

$$u^{(m)}(t) = u^{(m+1)}(t), \quad \forall t \in [0, t_m].$$

This implies that

$$\hat{u}^{(m)}(t) = \hat{u}^{m+1}(t), \quad \forall t \in [-t_m, 0].$$

Since m is arbitrary we conclude that if $u_0 \in \Lambda$, then (A.2) is uniquely solvable for $t < 0$.

♣

We have shown in chapter 4 that the Frechet derivative $Z(t, 0)$ of the completely continuous map $W(t, 0; \cdot)$ about a smooth basic time dependent solution is a compact linear operator. Thus, in particular, the Frechet derivative of $W(t, 0; \cdot)$ about each element of Λ is a compact linear operator. We can then apply a general theorem of Mallet-Paret[57] to deduce that the compact attractor Λ has finite Hausdorff dimension. Direct estimates of the dimension of Λ has been given by various authors[48, 8, 15].

The global attractor Λ however, need not be a manifold on which coordinate charts can be defined. This should certainly be the case for high Reynold's numbers unless the system is nongeneric in some sense (for example due to the presence of a symmetry group). Hence it will be very useful to find a *global invariant manifold* M_G of integer dimension such that $\Lambda \subset M_G$. Coordinates on such a manifold (atleast in principle) will provide us with an "exact Galerkin" method. A construction method for such a manifold (called an *inertial manifold*) can be formulated in the following way.

Let $\boldsymbol{P}_N$ be the orthogonal projection operator from $\boldsymbol{H}$ on to the invariant subspace of A correspond to the first N eigenvalues $\mu_1, \mu_2, \cdots, \mu_N$. Then we look for a manifold M_G defined by the graph of a map $\phi_G : \boldsymbol{P}_N \boldsymbol{H} \to (I - \boldsymbol{P}_N)\boldsymbol{H}$ with the following properties:

(i) ϕ_G and $D\phi_G$ are Lipschitz maps.

(ii) $M_G = graph\ \phi_G$ contains Λ.

(iii) M_G is invariant to the action of $W(t, 0; \cdot)$ in the neighborhood of Λ.

(iv) M_G is attractive.

It can then be shown that if the eigenvalues of the Stokes operator satisfy,

$$\mu_{N+1}^{1/2} \geq K_1 \text{ and } \mu_{N+1}^{1/2} - \mu_N^{1/2} \geq K_2, \forall N \geq N^*$$

then such manifolds exists $\forall N > N^*$. Such a theorem for Lipschitz manifolds (Lipschitz ϕ_G) was proved by Foias, Sell and Temam [22]. In [65], this theorem was improved to show that, if M_G exists then $D\phi_G$ will also be Lipchitz. Unfortunately the hypothesis required for the existence of such manifolds have not been verified even for the two dimensional Navier-Stokes problem. Hence the *existence theory for inertial manifolds remains as an open problem for hydrodynamics*. It is however, possible to establish the existence of such global invariant manifolds for certain regularizations of the Navier-Stokes equations[65].

Remark:

We recall here the invariant cone theorem proven in chapter 6 in the neighborhood of time periodic basic orbits. If such a theorem can be proven for arbitrary time dependent basic orbits then it would be possible to use another existence theorem for inertial manifolds given in [88].

Finally we will remark on the approximation and regularization methods to the Navier-Stokes system and their implications on the "structural stability" of the global attractor. Let $W^h(t, 0; \cdot)$ be a semigroup which approximates $W(t, 0; \cdot)$ in some sense. In [41], $W^h(t, 0; \cdot)$ was obtained using a Galerkin projection of the Navier-Stokes system to a finite dimensional space. In [66], $W^h(t, 0; \cdot)$ was the semigroup associated with a particular regularization of the Navier-Stokes equations. Suppose Λ_h denotes the global attractor associated with $W^h(t, 0; \cdot)$. Then, it has been shown in the above papers that

$$\Lambda_h \to \Lambda \text{ as } h \to 0$$

and Λ_h is upper semicontinuous at $h = 0$. Although these results have been proven only for two dimensional domains, their implication to computational fluid dynamics is profound.

Appendix B

Implications Of Group Action

The Navier-Stokes equations (in the absence of boundary conditions) are covariant to the action of the Euclidean group $\mathcal{E}(3)$. If the domain Ω as well as the boundary conditions are invariant to the action of $\mathcal{E}(3)$ or of its subgroup, then the coordinate representation on the unstable manifold has a special structure. In this section we will provide a brief analysis of this structure so that the invariant manifold theory (of hydrodynamic transition) in the presence of symmetry, will fit in the general analysis given by Ruelle[70]. As a first step let us provide a complete derivation of the covariance property of the Navier-Stokes equations with respect to the action a group $\mathcal{G} \subseteq \mathcal{E}(3)$. The action of $\mathcal{G}$ on $\boldsymbol{R}^3$ is characterized as follows. Let $\mathcal{L}_D(\boldsymbol{R}^3; \boldsymbol{R}^3)$ be the class of all linear diffeomorphisms from $\boldsymbol{R}^3 \to \boldsymbol{R}^3$. Then the group action $\mathcal{G} \to \mathcal{L}_D(\boldsymbol{R}^3; \boldsymbol{R}^3)$ is represented by $\psi_g(\boldsymbol{x})$ such that the homomorphism property

$$\psi_{g_1 g_2}(\boldsymbol{x}) = \psi_{g_1} \circ \psi_{g_2}(\boldsymbol{x}), \quad \forall g_1, g_2 \in \mathcal{G}$$

is satisfied.

Let us define the class of scalar fields V_s and vector fields V_v as

$$V_s = \{p; p : \boldsymbol{R}^3 \to \boldsymbol{R}\} \quad \text{and}$$

$$V_v = \{v; v : \boldsymbol{R}^3 \to \boldsymbol{R}^3\}$$

Let the induced representation of $\mathcal{G}$ in these spaces be denoted in the following way:

$\mathcal{G} \to \mathcal{L}_D(V_s; V_s)$ is τ_g and $\mathcal{G} \to \mathcal{L}_D(V_v; V_v)$ is T_g. Here τ_g and T_g satisfy the homomorphism property

$$\tau_{g_1 g_2} = \tau_{g_1} \circ \tau_{g_2}, \quad \forall g_1, g_2 \in \mathcal{G}$$

$$T_{g_1 g_2} = T_{g_1} \circ T_{g_2}, \quad \forall g_1, g_2 \in \mathcal{G}$$

We will now derive expressions for τ_g and T_g.

The pressure field $p(\boldsymbol{x})$ will be represented in the new coordinate system as $[\tau_g p](\boldsymbol{\psi}_g(\boldsymbol{x}))$. That is the action of $g \in \mathcal{G}$ induces

$$(\boldsymbol{x}, p(\boldsymbol{x})) \to (\boldsymbol{\psi}_g(\boldsymbol{x}), [\tau_g p](\boldsymbol{\psi}_g(\boldsymbol{x}))).$$

However, since the pressure field which is a scalar should be same in the two coordinate systems

$$[\tau_g p](\boldsymbol{\psi}_g(\boldsymbol{x})) = p(\boldsymbol{x}).$$

Since $\boldsymbol{\psi}(\cdot)$ is invertible, we can obtain the induced representation for the pressure field as

$$[\tau_g p](\boldsymbol{x}) = p(\boldsymbol{\psi}_g^{-1}(\boldsymbol{x})) \tag{B.1}$$

Let us denote $\boldsymbol{x} = \Phi(\boldsymbol{X}, t)$ as the particle trajectory. Then due to the action of $\mathcal{G}$, the point $\boldsymbol{x}$ is mapped to

$$\boldsymbol{\psi}_g(\boldsymbol{x}) = \boldsymbol{\psi}_g(\Phi(\boldsymbol{X}, t)).$$

Let us now consider the velocity

$$v(\boldsymbol{x}, t) = \frac{d}{dt} \Phi(\boldsymbol{X}, t) \quad \text{at } \boldsymbol{x} = \Phi(\boldsymbol{X}, t).$$

The group action induces

$$(\boldsymbol{x}, v(\boldsymbol{x}, t)) \to (\boldsymbol{\psi}_g(\boldsymbol{x}), [T_g v](\boldsymbol{\psi}_g(\boldsymbol{x}), t)).$$

Now,

$$[T_g v](\boldsymbol{\psi}_g(\boldsymbol{x}), t) = \frac{d\boldsymbol{\psi}_g(\Phi(\boldsymbol{X}, t))}{dt}$$

$$= D\boldsymbol{\psi}_g(\boldsymbol{x}) v(\boldsymbol{x}, t).$$

144

Thus

$$[T_g v](\boldsymbol{x}, t) = D\boldsymbol{\psi}_g(\boldsymbol{x}) v(\boldsymbol{\psi}_g^{-1}(\boldsymbol{x}), t). \tag{B.2}$$

Let us define the Navier-Stokes map $\mathcal{F} : V_v \times V_s \to V_v \times V_s$ with $\mathcal{F} = (\mathcal{F}_1, \mathcal{F}_2)$ defined as

$$\mathcal{F}_1(\boldsymbol{v}, p) = \boldsymbol{v}_t + (\boldsymbol{v} \cdot \nabla)\boldsymbol{v} + \nabla p - \nu \Delta \boldsymbol{v}$$

and

$$\mathcal{F}_2(\boldsymbol{v}, p) = \nabla \cdot \boldsymbol{v}$$

Let us denote by $\sigma_g = T_g \times \tau_g$ the induced representation $\mathcal{G} \to \mathcal{L}_D(V_v \times V_s; V_v \times V_s)$. Then

Theorem B.1 *The Navier-Stokes equations are covariant to the action of $\mathcal{G}$:*

$$\sigma_g \mathcal{F} = \mathcal{F} \sigma_g, \quad \forall g \in \mathcal{G} \subseteq \mathcal{E}(3).$$

Proof:

We need to show that

$$\mathcal{F}_1(T_g \boldsymbol{v}, \tau_g p) = T_g \mathcal{F}_1(\boldsymbol{v}, p) \quad \text{and}$$

$$\mathcal{F}_2(T_g \boldsymbol{v}, \tau_g p) = \tau_g \mathcal{F}_2(\boldsymbol{v}, p).$$

Consider $g \in \mathcal{E}(3)$. Then $\boldsymbol{\psi}_g(\cdot) = (\boldsymbol{a}, \mathcal{O})$ and

$$\boldsymbol{\psi}_g(\boldsymbol{x}) = \boldsymbol{a} + \mathcal{O}\boldsymbol{x}.$$

Here $\boldsymbol{a}$ is a vector and $\mathcal{O}$ is an orthogonal matrix: $\mathcal{O}^{-1} = \mathcal{O}^T$. We can check that

$$\boldsymbol{\psi}_g^{-1}(\boldsymbol{x}) = -\mathcal{O}^{-1}\boldsymbol{a} + \mathcal{O}^{-1}\boldsymbol{x}.$$

Let us use this result in (B.1) and (B.2) to get

$$[\tau_g p](\boldsymbol{x}) = p(\boldsymbol{y})$$

with $\boldsymbol{y} = \boldsymbol{\psi}_g^{-1}(\boldsymbol{x})$ and

$$[T_g \boldsymbol{v}](\boldsymbol{x}) = \mathcal{O}\boldsymbol{v}(\boldsymbol{y}).$$

We now consider $\mathcal{F}_2(T_g v, \tau_g p)$:

$$\nabla \cdot [T_g v](\boldsymbol{x}) = \frac{\partial}{\partial x^i}[T_g v]_i(\boldsymbol{x})$$

$$= \frac{\partial}{\partial x^i}(\mathcal{O}_{ij} v_j(\boldsymbol{y})) = \mathcal{O}_{ij}\frac{\partial v_j(\boldsymbol{y})}{\partial y^k}\frac{\partial y^k}{\partial x^i}.$$

Now, since

$$\frac{\partial y^k}{\partial x^i} = \mathcal{O}_{ik}$$

we get

$$\frac{\partial}{\partial x^i}[T_g v]_i(\boldsymbol{x}) = \mathcal{O}_{ij}\mathcal{O}_{ik}\frac{\partial v_j(\boldsymbol{y})}{\partial y^k}$$

$$= \frac{\partial v_i(\boldsymbol{y})}{\partial y^i} = \nabla \cdot \boldsymbol{v}(\boldsymbol{y})$$

since $\mathcal{O}^T \mathcal{O} = I$. Thus

$$\nabla \cdot [T_g v](\boldsymbol{x}) = [\tau_g(\nabla \cdot \boldsymbol{v})](\boldsymbol{x})$$

and hence we have proved that

$$\mathcal{F}_2(T_g v, \tau_g p) = \tau_g \mathcal{F}_2(\boldsymbol{v}, p).$$

We now consider $\mathcal{F}_1(T_g v, \tau_g p)$ which is the same as

$$\frac{\partial}{\partial t}[T_g v](\boldsymbol{x}) + [T_g v] \cdot \nabla[T_g v](\boldsymbol{x}) + \nabla[\tau_g p](\boldsymbol{x}) - \nu\Delta[T_g v](\boldsymbol{x}).$$

We have easily

$$\frac{\partial}{\partial t}[T_g v](\boldsymbol{x}) = T_g[\frac{\partial}{\partial t}v](\boldsymbol{x}).$$

Let us now consider the other terms. We have

$$\nabla_i[\tau_g p](\boldsymbol{x}) = \frac{\partial p(\boldsymbol{y})}{\partial x^i} = \frac{\partial p(\boldsymbol{y})}{\partial y^k}\frac{\partial y^k}{\partial x^i}.$$

Thus

$$\nabla_i[\tau_g p](\boldsymbol{x}) = \mathcal{O}_{ik}\frac{\partial p(\boldsymbol{y})}{\partial y^k} = T_g[\nabla p]_i(\boldsymbol{x}).$$

Hence

$$\nabla[\tau_g p](\boldsymbol{x}) = T_g[\nabla p](\boldsymbol{x}).$$

146

Now we consider $[T_g v] \cdot \nabla [T_g v](\boldsymbol{x})$, which is the same as

$$\mathcal{O}_{ik} v_k(\boldsymbol{y}) \frac{\partial}{\partial x^i}(\mathcal{O}_{jl} v_l(\boldsymbol{y})) = \mathcal{O}_{ik} \mathcal{O}_{jl} \mathcal{O}_{im} v_k(\boldsymbol{y}) \frac{\partial v_l(\boldsymbol{y})}{\partial y^m}$$

$$= \mathcal{O}_{jl} v_m(\boldsymbol{y}) \frac{\partial v_l(\boldsymbol{y})}{\partial y^m}.$$

Hence

$$[T_g v] \cdot \nabla [T_g v](\boldsymbol{x}) = [T_g(v \cdot \nabla v)](\boldsymbol{x}).$$

We have finally

$$[\Delta T_g v_i](\boldsymbol{x}) = \frac{\partial^2}{\partial x^j \partial x^j} \mathcal{O}_{ik} v_k(\boldsymbol{y})$$

$$= \mathcal{O}_{ik} \mathcal{O}_{jm} \mathcal{O}_{jl} \frac{\partial^2}{\partial y^m \partial y^l} v_k(\boldsymbol{y}) = \mathcal{O}_{ik} \frac{\partial^2}{\partial y^l \partial y^l} v_k(\boldsymbol{y}).$$

Hence

$$[\Delta T_g v](\boldsymbol{x}) = T_g [\Delta v](\boldsymbol{x}).$$

Collecting all these results we get

$$\mathcal{F}_1(T_g v, \tau_g p) = T_g \mathcal{F}_1(v, p).$$

We have thus proved the *covariance* $\sigma_g \mathcal{F} = \mathcal{F} \sigma_g$ of the Navier-Stokes equations under the action of $\mathcal{G}$.

♣

Definition B.1 *A open set* $\Omega \subset \boldsymbol{R}^3$ *is called a symmetry domain for* $\mathcal{G} \subset \mathcal{E}(3)$ *if*

$$\psi_g(\Omega) = \Omega.$$

Note that since $\mathcal{E}(3)$ is the group of rigid body rotations and translations, $\psi_g(\Omega) = \Omega$ implies that

$$\psi_g(\partial \Omega) = \partial \Omega.$$

Definition B.2 *A vector field* $u(\cdot) : \boldsymbol{R}^3 \to \boldsymbol{R}^3$ *is invariant to the action of* $\mathcal{G} \subset \mathcal{E}(3)$ *if*

$$[T_g u](\boldsymbol{x}) = u(\boldsymbol{x}), \quad \forall \boldsymbol{x} \in \boldsymbol{R}^3.$$

Let us now prove the covariance of the evolution form of the Navier-Stokes equations and the associated nonlinear semigroup $W(t,0;\cdot)$. We write

$$v_t = \mathcal{X}_u(v) \tag{B.3}$$

where

$$\mathcal{X}_u(v) = -Av - L_u v - B(v,v).$$

We will state

Theorem B.2 *Let $\Omega \subset \mathbf{R}^3$ be a symmetry domain for $\mathcal{G} \subset \mathcal{E}(3)$ and let the basic solution $U(\cdot,t) : \Omega \to \mathbf{R}^3$ is invariant to the action of $\mathcal{G}$. Then the evolution form of the Navier-Stokes equation is covariant:*

$$T_g \mathcal{X}_u = \mathcal{X}_u T_g, \quad \forall g \in \mathcal{G}$$

Proof:

Recall that the evolution form (B.3) can be obtained from the primitive form (2.1) by the projection $P_H : L^2(\Omega) \to \mathbf{H}$. Now, from the theorem B.1 and the invariance $[T_g U](\mathbf{x}, t) = U(\mathbf{x}, t)$ of the basic solution we conclude that the primitive form (2.1) is covariant to the action of $\mathcal{G}$. Hence, in order to deduce the commutive property of the evolution form, we only need to show that the projection operator P_H commutes with T_g:

$$P_H T_g = T_g P_H.$$

To show this we begin with the Hodge decomposition of a square integrable vector field $w : \Omega \to \mathbf{R}^3$,

$$w = v + \nabla \phi$$

Here ϕ is a scalar field and $v : \Omega \to \mathbf{R}^3$ satisfies

$$\nabla \cdot v = 0 \text{ and } v \cdot n|_{\partial\Omega} = 0.$$

That is $P_H w = v$. Note now that,

$$\nabla \cdot [T_g v](\mathbf{x}) = [\tau_g (\nabla \cdot v)](\mathbf{x}) = 0$$

148

and,

$$[T_g v](\boldsymbol{x}) \cdot \boldsymbol{n}(\boldsymbol{x})|_{\partial\Omega} = \mathcal{O}_{ij} v_j(\boldsymbol{y}) n_i(\boldsymbol{x})|_{\partial\Omega}$$

$$= v_j(\boldsymbol{y}) n_i(\boldsymbol{y})|_{\partial\Omega} = 0.$$

Here we have used the fact that $\psi_g(\partial\Omega) = \partial\Omega$. Hence

$$P_H[T_g w](\boldsymbol{x}) = [T_g v](\boldsymbol{x}) = [T_g(P_H w)](\boldsymbol{x})$$

That is $P_H T_g = T_g P_H$. This results immediately implies (by the proof of the theorem B.1) that

$$(i) \quad T_g A = A T_g \tag{B.4}$$

$$(ii) \quad T_g L_u = L_u T_g \tag{B.5}$$

$$(i) \quad T_g B(\cdot,\cdot) = B(T_g\cdot,T_g\cdot). \tag{B.6}$$

We thus have $\mathcal{X}_u T_g = T_g \mathcal{X}_u$.

♣

We will now prove a theorem concerning the covariance of various evolution operators used in the invariant manifold theory.

Theorem B.3 *Let $\Omega \subset \boldsymbol{R}^3$ be a symmetry domain for $\mathcal{G} \subset \mathcal{E}(3)$ and let the basic solution $\boldsymbol{U}(\cdot,t) : \Omega \to \boldsymbol{R}^3$ be invariant to the action of $\mathcal{G}$. Then $\forall g \in \mathcal{G}$ we have,*

$$(i) \quad T_g R(\lambda; -A) = R(\lambda; -A) T_g.$$

$$(ii) \quad T_g S(t) = S(t) T_g.$$

$$(iii) \quad T_g Z(T,0) = T_g Z(T,0) \quad and$$

$$(iv) \quad T_g W(T,0;\cdot) = W(T,0;T_g\cdot)$$

where $R(\lambda; -A)$, $S(t)$, $Z(T,0)$ and $W(T,0;\cdot)$ are respectively the resolvant operator, the Stokes semigroup, the monodromy operator and the nonlinear hydrodynamic semigroup.

Proof:

Since the Stokes operator is covariant we can write

$$(\lambda + A)T_g = T_g(\lambda + A).$$

Hence pre and post "multiplying" by $(\lambda + A)^{-1}$ we get

$$T_g R(\lambda; -A) = R(\lambda; -A)T_g.$$

Now recall that the Stokes semigroup generated by $-A$ can be represented by

$$S(t) = \frac{1}{2\pi i} \int_\Gamma e^{\lambda t} R(\lambda; -A) d\lambda$$

for an appropriate contour Γ. Hence the covariance of the resolvant operator implies that

$$S(t)T_g = T_g S(t), \quad t \geq 0.$$

In order to prove the covariance of the monodromy operator, we will consider the operator $\mathcal{K}$ introduced in chapter 4:

$$[\mathcal{K}v](t) = -\int_0^t S(t - \tau)L_u(\tau)v(\tau)d\tau.$$

Now

$$[\mathcal{K}T_g v](t) = -\int_0^t S(t - \tau)L_u(\tau)T_g v(\tau)d\tau$$

$$-\int_0^t T_g S(t - \tau)L_u(\tau)v(\tau)d\tau$$

$$= [T_g \mathcal{K}v](t)$$

due to the covariance of $S(\cdot)$ and L_u established earlier. Now, the evolution operator $Z(t,0)$ satisfies:

$$[I - \mathcal{K}]Z(t,0)v_0 = S(t,0)v_0.$$

Thus

$$T_g[I - \mathcal{K}]Z(t,0)v_0 = T_g S(t,0)v_0.$$

That is

$$[I - \mathcal{K}]T_g Z(t,0)v_0 = S(t,0)T_g v_0.$$

Hence

$$T_g Z(t,0)v_0 = \sum_{n=0}^{\infty} [\mathcal{K}^n S](t)T_g v_0$$
$$= Z(t,0)T_g v_0.$$

We thus have

$$T_g Z(t,0) = Z(t,0)T_g.$$

We finally come to the covariance of the nonlinear hydrodynamic semigroup. We have

$$T_g W(t,0;v_0) = T_g Z(t,0)v_0 - \int_0^t T_g Z(t,\tau)B(v(\tau),v(\tau))d\tau$$

with $W(t,0;v_0) = v(t)$. Using the covariance properties of $Z(t,0)$ and $B(\cdot,\cdot)$ we get

$$T_g W(t,0;v_0) = T_g v(t) = Z(t,0)T_g v_0 - \int_0^t Z(t,\tau)B(T_g v(\tau),T_g v(\tau))d\tau.$$

Now using the maps $\mathcal{F}$ and $\mathcal{M}$ introduced in chapter 5 we write this as

$$\mathcal{F}(T_g v, T_g v_0) = [T_g v](t) - Z(t,0)T_g v_0 + M(T_g v, T_g v) = 0. \tag{B.7}$$

Note that $\mathcal{F}(\cdot,\cdot)$ satisfies

$$T_g \mathcal{F}(v, v_0) = \mathcal{F}(T_g v, T_g v_0).$$

As in chapter 5 we use the implicit function theorem to solve (B.7) to get

$$[T_g v](t) = W(t,0;T_g v_0).$$

Hence

$$T_g W(t,0;v_0) = W(t,0;T_g v_0).$$

We have thus shown that the evolution equation $v_t = \mathcal{X}_u(v)$ as well as the associated nonlinear semigroup ("Semiflow") $W(t,0\cdot)$ are covariant to the action of $\mathcal{G}$.

$$\clubsuit$$

Let us now investigate the implications of the above covariance property of the Hydrodynamic semigroup on the structure of the unstable manifold. We will denote the restrictions of T_g to $P_u D(A^\alpha)$ and $P_s D(A^\alpha)$ by T_g^u and T_g^s respectively. The properties

$$T_g^u P_u = P_u T_g \quad \text{and}$$

$$T_g^s P_s = P_s T_g$$

follow from the definitions

$$P_u = \frac{1}{2\pi i} \int_{\Gamma_u} R(\lambda; Z(T,0)) d\lambda \text{ and}$$

$$P_s = \frac{1}{2\pi i} \int_{\Gamma_s} R(\lambda; Z(T,0)) d\lambda$$

and the result that ($Z(T,0)$ and hence) $R(\lambda; Z(T,0))$ commute with T_g.

Lemma B.1 *Let the graph of $\phi_u(\cdot) : P_u D(A^\alpha) \to P_s D(A^\alpha)$ describes the unstable manifold for $W(T,0;\cdot)$. Then*

$$T_g^s \phi_u(\cdot) = \phi_u(T_g^u \cdot).$$

Proof:

Let us begin with the same class of analytic maps $\phi(\cdot)$ used in chapter 6 with additional property $T_g^s \phi(\cdot) = \phi(T_g^u \cdot)$. We then have as before

$$\bar{x} = F_\phi(x) = P_u W(T,0; x + \phi(x))$$

and

$$[T_u \phi](\bar{x}) = P_s W(T,0; x + \phi(x)).$$

Note that

$$T_g^u F_\phi(x) = T_g^u P_u W(T,0; x + \phi(x))$$

$$= P_u T_g W(T,0; x + \phi(x)).$$

Now using the covariance of the semigroup operator we get

$$T_g^u F_\phi(x) = P_u W(T,0; T_g^u x + T_g^s \phi(x))$$

$$= P_u W(T, 0; T_g^u \boldsymbol{x} + \phi(T_g^u \boldsymbol{x})).$$

Thus

$$T_g^u F_\phi(\cdot) = F_\phi(T_g^u \cdot). \tag{B.8}$$

Similarly

$$T_g^s [T_u \phi](\bar{\boldsymbol{x}}) = T_g^s P_s W(T, 0; \boldsymbol{x} + \phi(\boldsymbol{x}))$$

$$= P_s W(T, 0; T_g^u \boldsymbol{x} + \phi(T_g^u \boldsymbol{x}))$$

$$= [T_u \phi](F_\phi(T_g^u \boldsymbol{x}) = [T_u \phi](T_g^u F_\phi(\boldsymbol{x}).$$

using (B.8). Thus

$$T_g^s [T_u \phi](\bar{\boldsymbol{x}}) = [T_u \phi](T_g^u \bar{\boldsymbol{x}}).$$

That is $T_g^s [T_u \phi](\cdot) = [T_u \phi](T_g^u \cdot)$.

We have thus shown that the construction of the unstable manifold can be performed in a T_g-invariant manner. Hence the fixed point argument used in chapter 6 will give us the existence of ϕ_u with the desired covariance property.

♣

These covariance properties of the maps can be used to simplify computations of the bifurcating solutions. In fact, if an analytic map commutes with T_g then, each multilinear operator involved in the (convergent) power series expansion of this map is covariant. This property can then be used to count the number of nonzero tensors associated with these multilinear operators. See the works of Sattinger[74] for a simple formula for the number of nonzero tensors in terms of Molien functions.

Bibliography

[1] R. A. Adams. *Sobolev Spaces*. Academic Press, New York, 1975.

[2] S. Agmon. *Lectures on elliptic boundary value problems*. Van Nostrand, New York, 1965.

[3] C. J. Amick. On the Leray problem of steady state Navier-Stokes flow past a body in the plane. *Acta Mathematica*, 161:71–130, 1988.

[4] C. J. Amick and L. E. Fraenkel. Steady solutions of the Navier-Stokes equations representing plane flow in channels of various types. *Acta Mathematica*, 144:84–152, 1980.

[5] K. I. Babenko. On stationary solutions of the problem of flow past a body of a viscous incompressible fluid. *Mat. USSR, Sbornik*, 91:3–26, 1973.

[6] K.I. Babenko. Periodic solutions of the flow of a viscous fluid past a body. *Sov.Math.Dokl*, 25:211–216, 1982.

[7] K.I. Babenko. Spectrum of the linearized problem of flow of a viscous incompressible liquid around a body. *Sov.Phy.Dokl*, 27:25–27, 1982.

[8] A.V. Babin and M.I. Vishik. Regular attractors of semigroups and evolution equations. *J. Math. Pure. Appl*, 62:441–491, 1983.

[9] J. Bemelmans. $C^{0+\alpha}$-Semigroups for flows past obstacles and for flows with capilary surfaces. In R. Rautmann, editor, *Approximation methods for Navier-Stokes problems*. Springer Verlag, 1980.

[10] M. E. Bogovskii. On the L^p theory of the Navier-Stokes system for unbounded domains with noncompact boundaries. *Soviet Math. Dokl*, 22:809–814, 1980.

[11] J.P. Bourguinon and H. Brezis. Remarks on the Euler equations. *Journal of functional analysis*, 15:341–363, 1974.

[12] L. Cattabriga. Su un problema al contorno relativo al sistema di equazioni di Stokes. *Rend. Mat. Sem. Univ. Padova*, 31:308–340, 1961.

[13] I-Dee Chang and R. Finn. On the solutions of a class of equations occuring in continuum mechanics, with applications to the Stokes paradox. *Arch. Rat. Mech. Analysis*, 7:388–401, 1961.

[14] E. A. Coddington and N. Levinson. *Theory of Ordinary Differential Equations.* McGraw-Hill, New York, 1955.

[15] P. Constantin, C. Foias, and R. Temam. Attractor representing turbulent flows. *Memoirs Amer. Math. Soc.*, 53(314), 1985.

[16] J.L. Daleckii and M.G.Krein. *Stability of solutions of differential equations in Banach space.* American mathematical society, Providence,R.I, 1974.

[17] G. Duvaut and J.L. Lions. *Inequalities in mechanics and physics.* Springer-Verlag, New York, 1976.

[18] M. A. Krasnosel'skii et al. *Approximate solutions of operator equations.* Moskou, 1969.

[19] R. Finn. On the steady-state solutions of the Navier-Stokes equations-III. *Acta Mathematica*, 105:197–244, 1961.

[20] R. Finn and W. Noll. On the uniqueness and nonexistence of Stokes flow. *Arch. Rat. Mech. Analysis*, 1:95–106, 1957.

[21] C. Foias and R.Temam. Remarques sur les equations de Navier-Stokes stationaires et les phenomenes successifs de bifurcation. *Annali Scuola Norm. Sup. Di Pisa*, V(1):29–63, 1978.

[22] C. Foias, G. R. Sell, and R. Temam. Inertial manifolds for nonlinear evolutionary equations. Preprint Series 234, IMA, University of Minnesota, March 1986.

[23] H. Fujita and T. Kato. On the Navier-Stokes initial value problem I. *Arch.Rat.Mech.Anal.*, 16(4):269–315, 1964.

[24] A. V. Fursikov. On some problems in control. *Dokl. Akad. Nauk SSSR*, pages 1066–1070, 1980.

[25] Y. Giga. Analyticity of the semigroup generated by the Stokes operator in L^r spaces. *Mathematische Zeitschrift*, 178:297–329, 1981.

[26] Y. Giga. Domains of fractional powers of the Stokes operator in L^r spaces. *Arch. for Rat. Mech. Analysis*, 89:251–265, 1985.

[27] Y. Giga and H. Sohr. On the Stokes operator in exterior domains. Preprint Series 29, Hokkaido University, May 1988.

[28] D. Gilbarg and H. F. Weinberger. Asymptotic properties of Leray's solution of the stationary two-dimensional Navier-Stokes equations. *Russian mathematical Surveys*, 29:109–123, 1974.

[29] D. Gilbarg and H. F. Weinberger. Asymptotic properties of steady state plane solutions of the Navier-Stokes equations with bounded Dirichlet integral. *Ann. Scuola Norm. Sup. Pisa*, 4:381–404, 1978.

[30] I.H. Herron. The Orr-Sommerfeld equation on infinite intervals. To appear in SIAM Review, 1989.

[31] J. G. Heywood. On some paradoxes concerning two dimensional Stokes flow past an obstacle. *Indiana Univ. Math. J.*, 24:443–450, 1974.

[32] J. G. Heywood. On the uniqueness questions in the theory of viscous flow. *Acta Math.*, 136:61–102, 1976.

[33] J. G. Heywood. The Navier-Stokes equations:on the existence, regularity and decay of solutions. *Indiana university mathematical journal*, 29(4):639–681, 1980.

[34] J. G. Heywood and S. S. Sritharan. The theory of time dependent viscous flow through two dimensional unbounded domains with noncompact boundaries. To be published., 1989.

[35] E. Hille and R.S.Phillips. *Functional analysis and semigroups*. 31. American mathematical society, Providence,R.I, 1957.

[36] M.W. Hirsch and C.C.Pugh. Stable manifolds and hyperbolic sets. In *Proceedings of the symposium in pure mathematics*. AMS, 1970.

[37] E. Hopf. Ein allgemeiner endlichkeitssatz der hydrodynamik. *Math. Ann.*, 117:764–775, 1941.

[38] E. Hopf. Uber die Aufangswertaufgabe fur die hydrodynamischen Grundgliechungen. *Math. Nachr.*, 4:213–231, 1951.

[39] G. Iooss. Estimation in the neighborhood of t=0, for an example of evolution problem where there is incompatibility between the boundary and initial data. *C.R.Acad.Sci.Paris*, 271:187–190, 1970.

[40] G. Iooss. Bifurcation of a periodic solution of the Navier-Stokes equations in to an invariant torus. *Arch. Rational Mech. Anal.*, 58:35–56, 1975.

[41] Xiao-Biao Lin J. K. Hale and G. Raugel. Upper semicontinuity of attractors for approximations of semigroups and partial differential equations. *Mathematics of Computations*, 50:89–123, 1988.

[42] L. V. Kapitanskii and K. I. Piletskas. On spaces of solenoidal vector fields and boundary value problems for the Navier-Stokes equations in domains with noncompact boundaries. *Pro. Steklev Inst. Math.*, 2:3–34, 1984.

[43] K. Kirchgassner and J.Scheurle. Bifurcation. In J.K.Hale L.Cesari and J.P.LaSalle, editors, *Dynamical systems*. Academic press, 1976.

[44] S. G. Krein. *Linear differential equations in Banach spaces*. American mathematical society, 1971.

[45] O. A. Ladyzhenskaya. *The mathematical theory of viscous incompressible flow*. Gordon and Breach, New York, second edition, 1969. English translation.

[46] O. A. Ladyzhenskaya. Mathematical analysis of Navier-Stokes equations for incompressible fluids. *Annual Rev. of Fluid Mechanics*, 7:249–272, 1975.

[47] O. A. Ladyzhenskaya. A dynamical system generated by the Navier-Stokes equations. *J. of Soviet Math*, 3:458–479, 1976.

[48] O. A. Ladyzhenskaya. On the finiteness of the dimension of bounded invariant sets for the Navier-Stokes equations and other related dissipative systems. In *Boundary value problems of mathematical physics and related questions in functional analysis*, volume 14. Steklev Institute, Leningrad, 1988.

[49] O.A. Ladyzhenskaya and V.A.Solonnikov. On the linearization principle and on invariant manifolds for problems of magnetohydrodynamics. *Zap.Nauch.Sem.Lomi. Leningrad*, 38:46–93, 1973.

[50] O.A. Ladyzhenskaya and V.A.Solonnikov. Some problems in vector analysis and generalized formulations of boundary value problems for the Navier-Stokes equations. *J. of Soviet Math*, 10:257–286, 1978.

[51] L. D. Landau. On the problem of turbulence. *C. R. Acad. Sci. U.R.S.S*, 44:311–314, 1944.

[52] J. Leray. Etude de diverses équations intégrales nonlinéaires et de quelques problémes que pose l'hydrodynamique. *J. Math. Pures Appl.*, 12:1–82, 1933.

[53] J. L. Lions. *Quelques méthodes de résolution des problémes aux limites non linéaires*. Dunod, Paris, 1969.

[54] J. L. Lions and E. Magenes. *Nonhomogeneous boundary value problems and applications*. Springer, Berlin, 1972.

[55] J. L. Lions and G. Prodi. Un théoréme d'existence et d'unicité dans les équations de Navier-Stokes en dimension 2. *C. R. Acad. Sci. Paris*, 248:3519–3521, 1959.

[56] J.L. Lions. Espaces d'interpolation et domaines de puissances fractionnaires d'operateurs. *J.Math.Soc.Japan*, 14(2):233–241, 1962.

[57] J. Mallet-Paret. Negatively invariant sets of compact maps and an extension of a theorem of Cartwright. *J. Differential equations*, 22:331–348, 1976.

[58] V. N. Maslennikova and M. E. Bogovskii. Sobolev spaces of solenoidal vector spaces. *Sibirian. Math*, 22:399–420, 1982.

[59] V. N. Maslennikova and M. E. Bogovskii. Approximation of potential and solenoidal vectorfields. *Siberian. Math*, 24:768–787, 1983.

[60] K. Masuda. On the stability of incompressible viscous fluid motions past objects. *J. Math. Soc. Japan*, 27(2):294–327, 1975.

[61] N. G. Meyers and J. Serrin. H=W. *Proc. Nat. Acad. Science. USA*, 51:1055–1056, 1964.

[62] H. K. Moffatt. The degree of knottedness of tangled vortex lines. *J. of Fluid Mechanics*, 35:117–129, 1969.

[63] A. C. Newell and J. A. Whitehead. Finite bandwidth, finite amplitude convection. *J. of Fluid mechanics*, 38:279–303, 1969.

[64] Y. R. Ou and S. S. Sritharan. Analysis of regularized Navier-Stokes equations-I. To be published, 1989.

[65] Y. R. Ou and S. S. Sritharan. Analysis of regularized Navier-Stokes equations-II. To be published. See also ICASE report 89-14, 1989.

[66] Y. R. Ou and S. S. Sritharan. Upper semicontinuous global attractors for viscous flow. To be published. See also ICASE Report 90-2, 1990.

[67] A. Pazy. *Semigroups of linear operators and applications to partial differential equations.* Springer-Verlag, New York, 1983.

[68] G. Prodi. Qualche risultato riguardo alle equazioni di Navier-Stokes nel caso bidimensionale. *Rend. Sem. Mat. Padova*, 30:1–15, 1960.

[69] G. Prodi. Teoremi di tipo lacale per il sistema de Navier-Stokes e stabilitá delle soluzione stazionarie. *Rend. Sem. Mat. Univ. Padova*, 32:374–397, 1962.

[70] D. Ruelle. Bifurcation in the presence of a symmetry group. *Arch. Rat. Mech. Analysis*, 51:136–152, 1973.

[71] D. Ruelle and F. Takens. On the nature of turbulence. *Comm. Math. Phys*, 20:167–192, 1971.

[72] B. V. Rusonov. Slow nonequilibrium flow of a viscous fluid past a cylinder. *Dokl. Akad. Nauk SSSR*, 89:983–986, 1953.

[73] D.H. Sattinger. The mathematical problem of hydrodynamic stability. *Journal of mathematics and mechanics*, 19(9):797–817, 1970.

[74] D.H. Sattinger. *Branching in the presence of symmetry.* SIAM, Pennsylvania, 1983.

[75] J. Scheurle. Selective iteration and applications. *Journal of Math.Analy. Applications*, 59:596–616, 1977.

[76] J. Serrin. Mathematical principles of classical fluid mechanics. In *Handbuck der physik.* Springer-Verlag, 1957.

[77] S. Smale. Differential dynamical systems. *Bull. Amer. Math. Soc.*, 73:747–817, 1967.

[78] P. Sobolevskii. On the nonstationary equations of the hydrodynamics of a viscous fluid. *Dokl. Akad. Nauk SSSR*, 128:45–48, 1959.

[79] V. A. Solonnikov. Estimates of the Green tensor for some boundary problems. *Dokl. Akad. Nauk. SSSR*, 130:988–991, 1960.

[80] V. A. Solonnikov. On the solvability of boundary and initial value problems for the Navier-Stokes system in domains with noncompact boundaries. *Pacific Journal of Mathematics*, 93:443–458, 1981.

[81] V. A. Solonnikov and K. I. Piletskas. Certain spaces of solenoidal vectors and solvability of boundary value problems for the Navier-Stokes system in domains with noncompact boundaries. *J. Sov. Math.*, 34:2101–2111, 1986.

[82] S. S. Sritharan. An optimal control problem in exterior hydrodynamics. In E. B. Lee G. Chen, L. Markus and W. Littman, editors, *New Trends And Applications Of Distributed Parameter Control Systems.* Marcel Dekker, 1990.

[83] G. G. Stokes. On the effects of the internal friction of fluids on the motion of pendulums. *Math. Phys. Papers*, 1851.

[84] J. T. Stuart. On the nonlinear mechanics of hydrodynamic stability. *J. of Fluid mechanics*, 4:1–21, 1958.

[85] H. Tanabe. *Equations of evolution.* Pitman, London, 1979.

[86] R. Temam. *Navier-Stokes Equations and nonlinear functional analysis.* CBMS-NSF Regional Conference Series in Applied Mathematics. SIAM, Philadelphia, 1983.

[87] R. Temam. *Navier-Stokes Equations.* North-Holland Publ. Comp., Amsterdam, third edition, 1984.

BIBLIOGRAPHY

[88] R. Temam. *Infinite dimensional dynamical systems in mechanics and physics.* Springer-Verlag, New York, 1988.

[89] M. J. Vishik and A. V. Fursikov. *Mathematical Problems in Statistical Hydromecanics.* Kluwer Academic publishers, Boston, 1988.

[90] I. I. Vorovich and V. I. Yudovich. Stationary flows of incompressible viscous fluids. *Mat. Sb.*, 53:393–428, 1961.

[91] W. Walter. *Differential and Integral Inequalities.* Springer-Verlag, Berlin, New York, 1970.

[92] J. Watson. On the nonlinear mechanics of wave disturbances in stable and unstable parallel flows hydrodynamic stability. *J. of Fluid mechanics*, 9:371–389, 1960.

[93] J.C. Wells. Invariant manifolds of nonlinear operators. *Pacific Journal of mathematics*, 62:285–293, 1976.

[94] V. I. Yudovich. Periodic solutions of viscous incompressible fluids. *Dokl. Akad. Nauk.SSSR*, 130:1214–1217, 1960.

[95] V. I. Yudovich. On the stability of forced oscilations of a liquid. *Dokl. Akad. Nauk.SSSR*, 195:292–295, 1970.

Index